Springer Proceedings in Complexity

Springer Proceedings in Complexity publishes proceedings from scholarly meetings on all topics relating to the interdisciplinary studies of complex systems science. Springer welcomes book ideas from authors. The series is indexed in Scopus.

Proposals must include the following:

- name, place and date of the scientific meeting
- a link to the committees (local organization, international advisors etc.)
- scientific description of the meeting
- list of invited/plenary speakers
- an estimate of the planned proceedings book parameters (number of pages/articles, requested number of bulk copies, submission deadline)

Submit your proposals to: Hisako.Niko@springer.com

More information about this series at https://link.springer.com/bookseries/11637

Georgi Yordanov Georgiev
Mahmoud Shokrollahi-Far
Editors

Efficiency in Complex Systems

Self-Organization Towards Increased Efficiency

Springer

Editors
Georgi Yordanov Georgiev
Physics Department
Assumption University
Worcester, MA, USA

Physics Department
Worcester Polytechnic Institute
Worcester, MA, USA

Mahmoud Shokrollahi-Far
University College of Nabi-Akram (UCNA)
Tabriz, Iran

Tilburg University
Tilburg, Noord-Brabant, The Netherlands

ISSN 2213-8684 ISSN 2213-8692 (electronic)
Springer Proceedings in Complexity
ISBN 978-3-030-69287-2 ISBN 978-3-030-69288-9 (eBook)
https://doi.org/10.1007/978-3-030-69288-9

This Springer imprint is published by the registered company Springer Nature Switzerland AG
The registered company address is: Gewerbestrasse 11, 6330 Cham, Switzerland

To my daughter, Victoria; my brother, Iskren; my wife, Boriana; my mother Velichka; and my father, Yordan!

Georgi Yordanov Georgiev

Acknowledgments

Special thanks to the Complex Systems Society, which made this meeting possible as one of its satellites at its 2017 Annual Conference in Cancun, Mexico. Thank you also to all presenters at the meeting and contributors to this volume, with which their insights advance the search for the Grand Unification of all processes and structures in nature. Thanks go to Springer for the enthusiasm to publish on such cutting-edge topics. Thanks to the reader, without whom this book will be useless.

Introduction

Complexity has been rediscovered in the search for simplicity. The reductionism, leading to the search for smaller and smaller building blocks of nature, has revealed that the subatomic world is as emergent and unpredictable as complexity at the largest of scales. This provides the ground for a grand unification of quantum mechanics, and eventually quantum gravity, with complexity theory. Both of them are statistical and describe systems that are self-organizing and evolving. Quantum entanglement points to structures and organization in systems correlated on long distances. The Copenhagen interpretation of quantum mechanics points to a picture where the elementary particles are emergent from the subatomic energy fluctuations by interactions. In macroscopic systems, organisms, ecosystems, economies, languages, and engineered systems emerge similarly from the fluctuations of the energy in the primordial state of their constituent elements in mutual interaction.

This book is on the cusp of this grand unification of science, a path towards a true theory of everything, starting from the Planck's scale to the scale of the Universe, and all of the complex systems in between, such as atoms, molecules, cells, organisms, economies, technologies, civilization, information, consciousness, and all else that we see and will see around us. A theory will be of everything if it explains not just the interaction of subatomic particles, black holes, and the big bang, but also the existence of the all else that we can observe, perceive, and understand, including us, our thoughts, intelligence, and consciousness, all in one breath, and may be with one equation.

Why do we think that efficiency underlies all of those processes, and which efficiency are we talking about? All events in nature occur with the least expenditure of action, in terms of energy and time. This makes them most action efficient. That is how all of the physical laws are derived. Even when longer trajectories are allowed and possible, the most action efficient ones are those that are predominant and the structure of systems is set in such a way as to ensure action efficiency, i.e., to obey the natural laws. Conceptually, action efficiency is that the most possible is the most probable. If a process occurs with lower efficiency, it consumes more resources, and when those are scarce, it cannot occur. The processes which occur most often are the ones that have the greatest chance for availability of those resources, and therefore they are the ones possible and most frequent. The more inefficient a process is, the more impossible it becomes. This is the nature of the principle of least action. There is nothing mysterious about it. And, as a first principle, it applies to everything, from the vacuum fluctuations and the smallest building blocks of the universe, to the largest possible systems in it. It is one of the true grand unification principles on all scales. It guides the behavior of all systems.

We hope that the articles in this volume will spark further work in this journey, discussions, connections, collaborations, and discoveries that will pave the way for a more complete understanding of nature. This is the largest dream that all of us had since the first spark of consciousness, since the first time humans had looked at the stars and wondered what they are and who we are—a single underlying explanation for all that exists and that will exist.

Contents

On the Origin of Universal Patterns

Arto Annila

Abstract The most comprehensive result of scientific inquiry across disciplines is that data, irrespective of origin, display skewed distributions, sigmoid curves, and power laws as well as oscillations and, at times, chaos. While mathematical models and computer simulations can be made to reproduce these ubiquitous patterns of nature, science is not only about modeling and mimicking the data but making sense of it. We argue that the ubiquitous patterns follow from the least-time consumption of free energy. These natural processes can be described by the many-body theory of open systems, i.e., nonequilibrium statistical physics for quantized systems. This theory, also known as the second law of thermodynamics, explains the arrow of time in terms of flows of quanta as well as non-determinate and path-dependent evolution that yields the scale-free patterns.

Keywords Atomism · Complexity · Free energy · The principle of least action · Scale-free patterns · The second law of thermodynamics

1 Introduction

Today, the spectrum of scientific knowledge extends from tiny elementary particles to gigantic galaxies and from the richness of genes to the abundance of species. As startling as it is, the data are highly similar, regardless of what we look at. The universal characteristics are evident in immense masses of information called "big data" (Albert and Barabási 2002; Clauset et al. 2009; Newman 2005; Bak 1997; Sornette 2006; Buchanan 2002). The mathematical models of lognormal distributions, S-curves, and power laws match data irrespective of the field. Complex

A. Annila (✉)
Department of Physics, FI-00014 University of Helsinki, Helsinki, Finland

Department of Biosciences, FI-00014 University of Helsinki, Helsinki, Finland
e-mail: arto.annila@helsinki.fi

G. Y. Georgiev, M. Shokrollahi-Far (eds.), *Efficiency in Complex Systems*, Springer Proceedings in Complexity, https://doi.org/10.1007/978-3-030-69288-9_1

systems science is the new discipline that models this great regularity (Waldrop 1993).

Unless headers and units are labeled in the descriptors of different datasets, we could not say from where the data originates. For example, the length distribution of genes looks much like the length distribution of words. The distributions of animal and plant populations are just as skewed as the distributions of genes and words. Distributions of wages and wealth are also skewed. The size distribution of earthquakes looks similar to the size distribution of the activated cortical areas in the brain (Bak 1997). Moreover, natural spirals, such as clamshells, heads of flowers, hurricanes, and galaxies, all whirl in the same way (Hargittai and Pickover 1992; McNally 2010). Many growth curves follow the characteristic form of the letter "S"; that is, they are sigmoidal. For example, a bacterial population grows thus. Chemical reactions progress and economies develop likewise. The world is clearly not random but regular. Could it be consistent with just one single rule?

At first, it may seem all crazy to compare any data to any other data without any unit of measure. But, in this way, we are free from barriers to realizing that the world is everywhere amazingly similar. Only the units and scales that we have set ourselves vary from one dataset to another. When we cannot infer the origin of data from the data itself, we must accept that the data are similar, although not identical. The regularity stands for that which we cannot distinguish one dataset from another. What does this point to?

The similarity of the data is inconceivably broad. It is expressly puzzling unless we can see a common cause. The more general the explanation we must look for, the more different things share the same shape. Also, Newton was after the same reason for similar natural phenomena in his rules of scientific reasoning (Rossi 2001). The great regularity has been noticed. It has been modeled but not explained. For example, the lognormal distribution is a good model of skewed distributions. However, it does not say why the data distributes nearly lognormally. Likewise, the logistic curve matches many datasets of growth. However, a good fit does not say why growth is sigmoidal.

Statistical mechanics, as a theory of many-body systems, has the potential to explain the origin of the universal patterns. That promise is realized with statistical physics of open systems. According to that theory, flows of quanta between the system and its surroundings drive toward the mutual thermodynamic balance in the least time. The analysis reveals that evolution results in skewed distributions, sigmoid growth curves, spirals, and power laws that are found throughout nature.

2 Statistical Physics of Open Systems

At one time, Ludwig Boltzmann understood that gas attains thermodynamic balance by way of the gas molecules colliding on each other and on walls of the tank. Albert Einstein understood conversely that Planck's law of radiation accounts for light as a gas of photons. Willard Gibbs, in turn, comprehended that chemical compounds attain thermodynamic balance via reactions. Now, we can understand likewise the quanta, e.g., photons, as the fundamental elements of everything, redistribute energetically ever more favorably in all kinds of events. This process toward thermodynamic balance can be described by the equation of evolution, which, in turn, can be derived from the equation of state.

2.1 The Equation of State

The equation of state describes any system that comprises the basic building blocks, the quanta. In events, the system moves from one state to another either by gaining quanta from the surroundings or by giving away quanta to the surroundings. When all systems consist of quanta, the description is inherently universal. Thus, we can start by examining any constituent of any system. A constituent, indexed with j, exists with the probability ${}^1P_j = \phi_1\phi_2\phi_3 \ldots = \Pi_k\phi_k$, which is the product, Π_k, of its ingredients, indexed with k. Thus, if any one entity k is missing altogether, i.e., $\phi_k = 0$, then also ${}^1P_j = 0$. For example, an enzyme could not exist if any one of its ingredients was missing.

The ingenuity of statistical physics is that we can express the probability ${}^1P_j$, even when we do not know what entities ϕ_k are in the product Π_k because ultimately all entities comprise the quanta. Therefore, the equation of state includes all the details with the formal precision of the quantum.

When the system houses several entities of equal energy, for example, a cell having multiple copies of the same enzyme, the probability of the population $P_j = [{}^1P_j][{}^1P_j][{}^1P_j] \ldots /N_j! = [{}^1P_j]^N{}_j/N_j!$ is the product of the partial probabilities, where N_j is the size of the population. The product form ensures that if any one entity is missing, i.e., ${}^1P_j = 0$, then also $P_j = 0$. When the entities are identical, their mutual order makes no difference. Hence, the expression of P_j is divided by the number of ways $N_j!$ that the entities, in total N_j, can be arranged into a sequence (Fig. 1).

The total probability P of the system is the product Π_j of the partial probabilities P_j:

$$P = \prod_{j=1} P_j = \prod_{j=1} \left[\prod_{k=1} \phi_k \right]^{N_j} / N_j! \qquad (1)$$

where each factor $\phi_k = N_k \exp[(-\Delta G_{jk} + i\Delta Q_{jk})/k_BT]$ denotes the number of starting materials N_k and the energy difference ΔG_{jk} between the starting material, indexed with k, and the product, indexed with j. The higher the energy of the starting materials, the less energy $i\Delta Q_{jk}$ is needed from the surroundings to bridge the energy gaps from the starting materials into the product. The label i in front of the energy term means that the system is open for the flows of quanta. For example, the

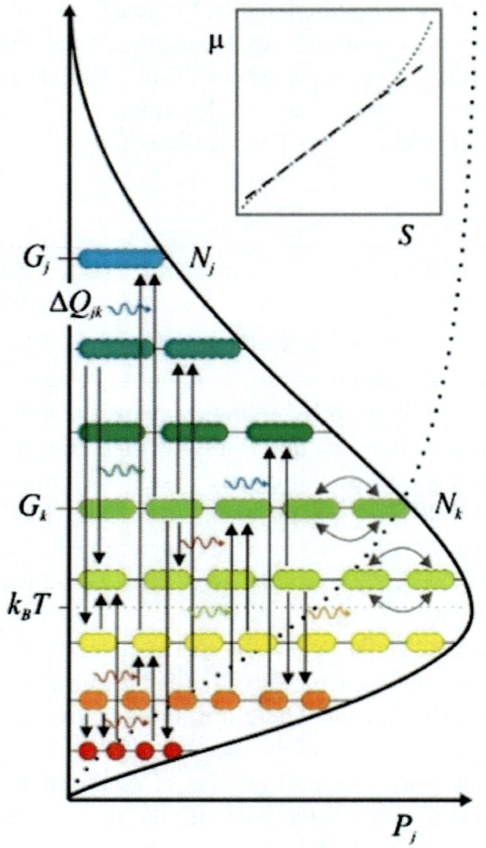

Fig. 1 When everything comprises quanta, any system can be pictured in terms of an energy-level diagram. The entities of a system, in numbers N_k, which have the same energy G_k, are on the same level. The bow arrows indicate their mutual exchange, which changes nothing and hence causes no change in the average energy of the system k_BT either. By contrast, the vertical arrows indicate events, in which the entities move from one level to another. For example, in a chemical reaction, starting materials N_k transform into products N_j. The horizontal wave arrows denote the quanta of light that either come from the environment to the system or go away from the system to the environment. Since the quanta carry energy ΔQ_{jk}, the events as flows of quanta move the system and its surroundings toward the thermodynamic balance. When the energy of the surroundings is higher than that of the system, the system will evolve toward higher average energy and the surrounding systems toward lower average energy, and vice versa. The cumulative probability distribution curve (dotted line) is a sigmoid. When the logarithm of the total probability, i.e., entropy S, is plotted as a function of [chemical] potential energy μ, the S-shaped curve follows on the logarithm-logarithm scale mainly a straight line (inset), that is, it follows a power law

photons from the Sun make photosynthesis happen. Conversely, the heat generated by our body goes away into the surroundings.

Energy differences between the products and starting materials are relative to the average energy of the system, k_BT. Since the concept of temperature T was adopted long before the concept of energy, T is multiplied by the Boltzmann constant k_B to make it commensurate with the other terms of energy. Naturally, the average energy changes in every event. However, when a single event perturbs the average energy only a little bit, the system evolves smoothly.

The state equation (Eq. 1) is the theory in essence. Thus, it is relevant to summarize the assumptions that have already been made since conclusions just follow from a straightforward mathematical analysis. (1) The state equation applies when the same elements make up everything. This atomistic axiom underlies statistical mechanics in general. (2) The system is statistical when there are numerous entities. Then the average energy $k_B T$ is a meaningful concept, and the energy differences relative to it can be expressed as exponential functions (exp) (Gibbs 1948; Phillies 2017). When these assumptions hold, the statistical theory explains why data of various kinds have the same form.

The state of the system is customarily given by an additive measure, which is obtained as the logarithm (ln) of the state equation (Eq. 1). For historical reasons, the logarithm of probability, when multiplied by the Boltzmann constant k_B, is known as the entropy:

$$S = k_B \ln P = k_B \sum_{j=1} \ln P_j \approx \frac{1}{T} \sum_{j=1} N_j \left(\sum_{k=1} -\Delta\mu_{jk} + i\Delta Q_{jk} + k_B T \right). \tag{2}$$

In the equation, $\Delta\mu_{jk}$ denotes the potential energy difference between the populations N_k and N_j. The energy that is bound to the k-entity population $\mu_k = k_B T \ln\phi_k$ is called [chemical] potential. Similarly, μ_j denotes the potential energy of the j-entities. In the equation for entropy (Eq. 2), the entry $\approx$ stands for the statistical approximation $\ln N_j! \approx N_j \ln N_j - N_j$, which is excellent when $N_j > 10$. It is worth emphasizing that the entropy expression (Eq. 2) is just the logarithm of probability (Eq. 1). In other words, mathematics does not change anything. It just keeps conclusions within the atomistic axiom of the theory.

2.2 *The Equation of Evolution*

The total energy of the system TS equals temperature T times entropy S. It comprises the system-bound energy $\Sigma N_j k_B T$ and the free energy $\Sigma N_j(-\Delta\mu_{jk} + i\Delta Q_{jk})$. When free energy $-\Delta\mu_{jk} + i\Delta Q_{jk}$ is decreasing, the populations N_j are changing, and hence, also the total energy of the system TS (Eq. 1) is changing with time t:

$$T\frac{dS}{dt} = \sum_{j=1} \frac{dS}{dN_j}\frac{dN_j}{dt} = \sum_{j=1} \frac{dN_j}{dt} \left(\sum_{k=1} -\Delta\mu_{jk} + i\Delta Q_{jk} \right). \tag{3}$$

It is convenient to denote the change as continuous, i.e., as a differential dN_j because for a statistical system, the change, quantum by quantum, appears as if it were continuous.

We cannot solve the equation of motion (Eq. 3) because the change in population is proportional to free energy, i.e., force:

$$\frac{dN_j}{dt} = \frac{1}{k_B T} \sum_{k=1} \sigma_{jk} \left(-\Delta\mu_{jk} + i\Delta Q_{jk}\right), \quad (4)$$

where $\sigma_{jk} > 0$ represents a mechanism that facilitates the flow of quanta. For example, an enzyme catalyzes the conversion of starting materials N_k into the products N_j or vice versa. The flows of quanta naturally select efficient mechanisms because then the thermodynamic balance is attained in the least time.

While the course of events cannot be predicted because forces and flows cannot be separated, the process can be simulated step by step, according to Eq. (4). In this way, the emergence of standards, skewed divisions, growth curves, oscillations, and chaotic courses can be demonstrated and modeled (Annila and Annila 2008; Jaakkola et al. 2008a, 2008b; Karnani and Annila 2009; Annila and Salthe 2009).

Equations (3) and (4) describe the flows of quanta so that the imbalance between the system and the environment decreases in the least time. When we substitute in Eq. (3) the change in the number dN_j/dt with Eq. (4), we see that the entropy cannot decrease. This is known as the second law of thermodynamics $dS \geq 0$. There is no exception since the quanta, as conserved entities, cannot come out of nothingness or vanish into nothingness. The quantum that leaves the system will end up in the environment or vice versa.

According to Eqs. (3) and (4), there are no energy barriers for the evolution from one state to another. If such barriers existed, thermodynamics and kinetics would be in conflict with each other. This is not the case. Free energy can only decrease. For example, the flow does not open until the water level rises over the spillway crest. Likewise, the chemical reaction does not proceed from the starting material to the product until the energy of the starting materials, including absorbed photons, exceeds the energy of the products. The catalyst does not change the energy level diagram or landscape. It is a mechanism that only speeds up the conversion of the starting materials into the products or vice versa. According to the theory of time, the flows direct so that energy differences are diminishing as soon as possible. Thus, entropy does not just increase, but it increases as fast as possible.

3 The Universal Patterns

The equation of evolution (Eq. 4) reproduces the S-shape of growth piecewise. At the beginning of the growth, there is a wealth of resources, i.e., free energy. Then, we can assume that mechanisms $\Sigma_k \sigma_{jk}$ of the system consume free energy $-\Delta\mu_{jk} + i\Delta Q_{jk}$ almost steadily, and hence the population N_j changes with time:

$$\frac{d}{dt}\frac{1}{k_B T}\sum_{k=1}(-\Delta\mu_{jk} + i\Delta Q_{jk}) = \frac{dN_j}{dt}\frac{d}{dN_j}\frac{1}{k_B T}\sum_{k=1}(-\Delta\mu_{jk} + i\Delta Q_{jk}) \approx \sum_{k=1}\sigma_{jk}$$
$$\Rightarrow \frac{dN_j}{N_j} = \sum_{k=1}\sigma_{jk}dt. \tag{5}$$

Here $d\mu_j/dN_j = k_B T/N_j$. The growth by Eq. (5) is approximately exponential because initially, the amount of free energy seems as if it were infinite, and only the mechanisms are limiting the growth. Likewise, the growth is decreasing almost exponentially when the free energy is dwindling down while the balance is approached.

The growth between the initial and final phases follows a power law closely. We see this by expressing the population N_j as the product of the elements N_1 using the atomistic axiom $N_j = \Pi_k \phi_k = \alpha_j N_{1j}$. The factor $\alpha_j = \Pi_{\text{mn}}\exp[\Sigma N_j(-\Delta\mu_{mn} + i\Delta Q_{mn})/k_B T]$ contains the free energy terms that force the assembly of N_j from the elements N_1. So, the change

$$\frac{dN_j}{dt} = j\alpha_j N_1^{j-1}\frac{dN_1}{dt} = j\frac{N_j}{N_1}\frac{dN_1}{dt} \quad \Rightarrow \quad \frac{dN_j}{N_j} = j\frac{dN_1}{N_1} \tag{6}$$

when integrated follows the power law $\ln N_j = j\ln N_1 +$ constant.

When the assumption of a nearly constant change in free energy does not hold, we can *model* the change by adding the term $-\beta N_j$ to the equation of the population change (Eq. 5):

$$\frac{dN_j}{N_j} \approx \left(\sum_{k=1}\sigma_{jk} - \beta N_j\right)dt \quad \Rightarrow \quad N_j(t) = N_j(t_0)\left(\sum_{k=1}\sigma_{jk} - \beta N_j(t_0)\right). \tag{7}$$

In this model (Mäkelä and Annila 2010; May 1976), the population $N_j(t_0)$ at present t_0 determines the population $N_j(t)$ at a later time t. According to the model, evolution is almost predictable when the change in free energy is small compared with average energy, i.e., $|(-\Delta\mu_{jk} + i\Delta Q_{jk})/k_B T| << 1$. In contrast, oscillations and chaos occur when the condition is not fulfilled. This is the case, for example, when a solid-state laser is turned on or when the animal population proliferates and exceeds the carrying capacity of the environment or when the banks need more money than is available.

Logarithmic, exponential, or truncated distributions and their power-law-like cumulative distribution functions are mathematical models of the physical processes given by Eqs. (3) and (4). The models allow us to describe and categorize various data, but they do not explain how the data came about, that is, causality. It is worth emphasizing that Eqs. (3) and (4) cannot be solved, except at balance, since the variables cannot be separated. This means that a chain of events is fundamentally unpredictable due to mutual dependencies rather than due to complexity or uncertainty in the initial conditions.

When the system evolves gradually, the change in energy is small compared with the average energy, i.e., $|(-\Delta\mu_{jk} + i\Delta Q_{jk})/k_BT| << 1$. Therefore, the variation n is small $n << j$ around a typical or an average factor, indexed with j. When the factors ϕ_j are given in logarithmic terms $\ln\phi_j = j\ln\phi_1$ of the elemental factor ϕ_1, we see that the natural distribution

$$\ln\phi_{j-n,j+n,j,n} = \ln\phi_j + \sum\nolimits_n n\ln\phi_1 \tag{8}$$

is approximately logarithmic. The distribution of Eq. (8) shows that the typical form j can be recognized in each member within the distribution $j \pm n$. For example, all-sized Northern pike looks like pike and not bream. On the other hand, if weights of pikes and cars were presented in the same figure, we would see two distributions: one about a typical pike and the other about an ordinary car. Moreover, spiral forms of nature, such as shells, cyclones, and galaxies, are also approximately lognormal distributions but in polar coordinates (Mäkelä and Annila 2010). Logarithmic spirals are thus energetically optimal shapes.

4 Discussion

Statistical physics of open systems accounts for processes as flows of quanta in accordance with observations and measurements. The correspondence between the theory and data implies further that everything comprises the indivisible elements, quanta, and every process seeks thermodynamic balance in the least time. The natural law is contained in the quantum itself. Planck's constant $h = Et$, as the complementary product of energy and time, determines the change in energy over time $dE/dt = -E/t = -\mathbf{F}\cdot\mathbf{v}$. Thus, the power dE/dt decreases due to motion with velocity $\mathbf{v}$ in the direction of force $\mathbf{F}$.

From this perspective, statistical physics of open quantized systems could be falsified (i) if a phenomenon were found where the system moves away from the thermodynamic balance; (ii) if a quantum, say, a photon, was found to split into pieces; or (iii) if something were found that is not quanta. Earlier, we have argued that elementary particles, as well as the void, comprise quanta (Annila 2012; Grahn et al. 2018; Annila 2010; Annila and Kolehmainen 2016).

5 Conclusions

Traditional statistical mechanics is limited to closed or equilibrium systems. When no net fluxes are included, the system is stationary, and hence calculations are precise. In contrast, the statistical physics of open systems, i.e., nonequilibrium statistical mechanics, includes net fluxes, and accordingly, the system is described in evolution from one state to another. However, the evolution is nondeterministic because the boundary conditions, the surroundings, which are the sources and sinks of the fluxes, are changing too. The future is genuinely unpredictable.

References

Albert R, Barabási A-L (2002) Statistical mechanics of complex networks. Rev Mod Phys 74:47

Annila A (2010) All in action. Entropy 12:2333

Annila A (2012) The meaning of mass. Int J Theor Math Phys 2:67

Annila A, Annila E (2008) Why did life emerge? Int J Astrobio 7:293

Annila A, Kolehmainen E (2016) Atomism revisited. Phys Essays 29:532

Annila A, Salthe S (2009) Economies evolve by energy dispersal. Entropy 11:606

Bak P (1997) How nature works. Oxford University Press, Oxford

Buchanan M (2002) Ubiquity: why catastrophes happen. Three Rivers Press, New York

Clauset A, Shalizi CR, Newman MEJ (2009) Power-law distributions in empirical data. SIAM Rev 51:661

Gibbs JW (1948) The collected works of J. Willard Gibbs. Yale University Press, New Haven

Grahn P, Annila A, Kolehmainen E (2018) On the carrier of inertia. AIP Adv 8:035028

Hargittai I, Pickover CA (1992) Spiral symmetry. World Scientific, Singapore

Jaakkola S, Sharma V, Annila A (2008a) Cause of chirality consensus. Curr Chem Biol 2(53):arXiv:0906.0254

Jaakkola S, El-Showk S, Annila A (2008b) The driving force behind genomic diversity. Biophys Chem 134(136):232

Karnani M, Annila A (2009) Gaia Again Biosyst 95:82

Mäkelä T, Annila A (2010) Natural patterns of energy dispersal. Phys Life Rev 7:477

May RM (1976) Simple mathematical models with very complicated dynamics. Nature 261:459

McNally J (2010) Earth's Most stunning natural fractal patterns. Science

Newman MEJ (2005) Power laws, Pareto distributions and Zipf's law. Contemp Phys 46:323

Phillies, GDJ (2017) Readings and misreadings of J. Willard Gibbs elementary principles in statistical mechanics. arXiv:1706.01500

Rossi P (2001) The birth of modern science. Wiley, Hoboken, pp a:42–55, b:63–66, c:140, d:170, e:206, f:213–214, g:237, h:264, i:297–300, j:325–328.

Sornette D (2006) Critical phenomena in natural sciences: chaos, fractals, self-organization and disorder: concepts and tools. Springer, Cham

Waldrop MM (1993) Complexity: the emerging science at the edge of order and chaos. Simon and Schuster, New York

Natural Classes and Natural Classification

Lauri Lehmonen and Arto Annila

Abstract Categorization is a natural way for us humans to differentiate one object from another as well as to relate entities to each other. However, are there classes in nature independent of human categorization? And is there a fundamental way of classification free from human cataloging? We consider that all objects can be categorized based on their ultimate composition of elemental building blocks, quanta. Our conjecture parallels that of Noether's theorem but follows from statistical mechanics of open systems. We conclude that the natural categorization places objects to classes so that free energy is consumed in the least time. While the imperative is universal, any classification is subjective. We relate these resolutions to conventional methods of categorization.

Keywords Dissipation · Entropy · Free energy · Photon · The principle of least action · The second law of thermodynamics

1 Introduction

Are categories truly natural notions or only conceptual constructs? We, humans, tend to be so consumed in categorizing perceptions that we hardly attend to our category-making. What is the basis of our categorization, and why do we place objects in categories in the first place? In fact, often we presume that there would be distinct categories, for instance, as antonyms, when asking fundamental questions "What is life?" and "What is consciousness?" Perhaps resolutions to these profound

L. Lehmonen
Department of Physics, FI-00014 University of Helsinki, Helsinki, Finland

A. Annila (✉)
Department of Physics, FI-00014 University of Helsinki, Helsinki, Finland

Department of Biosciences, FI-00014 University of Helsinki, Helsinki, Finland
e-mail: arto.annila@helsinki.fi

G. Y. Georgiev, M. Shokrollahi-Far (eds.), *Efficiency in Complex Systems*,
Springer Proceedings in Complexity, https://doi.org/10.1007/978-3-030-69288-9_2

questions (Sharma and Annila 2007; Annila 2016a) and others will first follow from thorough comprehension about classes and classification.

Aristotle's categorization of objects by successively narrowing questions such as "Is it animate or inanimate?" logically implies that one entity can be ultimately distinguished from another by an indivisible constituent. The ancient as well as modern atomism (Palmer 2012; Berryman 2008; Chalmers 2014) claims that everything comprises undividable basic building blocks. Indeed, humankind has progressed in making ever-finer distinctions manifesting today, e.g., as DNA-based taxonomy and lineages of elementary particles.

In terms of modern physics, the ultimate unit of increment is one quantum of action (Heisenberg 1927). The uncertainty principle excludes from categorizing any observation below the exactness of the quantum because the observation process itself will change its target at least by one quantum (Jordan 1934). In this sense, a natural class is defined by the basic entity whose properties are set (Dretske 1977).

This atomistic view was posited by Lewis in 1926 based on Planck's discovery of a constant and Einstein's interpretation of it as the quantum of light as well as by Noether's theorem that equates the number of quanta with system's energy and characteristic period of motion. We acknowledge that this old quantum theory was largely set aside when physics moved to modern quantum theory. Nonetheless, we are motivated to adopt the old concept and propose anew that the quantum of action is the basic constituent. This thesis means that a single quantum is the "natural" unit of classification. In the following, we will consider the consequences of this conjecture.

Although the single quantum can be regarded as the ultimate resolution of any object, many a categorization does not focus on the number of constituents but on functional differences among objects. The spectrum of functions in *scala naturae* is undoubtedly broad, but in terms of physics, all functions are alike. Namely, any process is some flow of energy (Sharma and Annila 2007; Du Châtelet 1759; De Maupertuis 1746; Mäkelä and Annila 2010). Thus, whether one entity can be distinguished from another by function depends on the subject's ability to discern differences in the flows of energy between one class of entities and another. In other words, there is no objective way of categorizing. Still, the flows of energy are not arbitrary either but comply with thermodynamics.

These preliminaries on the ultimate resolution and subjective character of classification imply on the one hand that there are natural categories distinct from each other by the number of quanta and on the other hand that objective and universal standards for categorization, albeit desired, are elusive. We motivate this insight by formulating a theory of classes and classification from the profound principles.

2 Hypotheses

The emergence of new classes and evolution to greater hierarchy are typical but not exclusive processes to biological systems (Salthe 1993; Ulanowicz and

Hannon 1987). Clearly, animate can distinguish, for instance, edible from harmful. Technological progress is characterized by ever-finer distinctions. The increasing competence in delineation is also reflected in increasing vocabulary. Specialized terminology meets specific needs. For instance, Sami languages in northern Europe have a wealth of snow- and ice-related words. Today English expands with words related to information technology. Applications of artificial intelligence, so-called expert systems (Jackson 1998), also depend on the ability to classify "correctly."

At this point, it is worth recalling that in philosophy, it has been debated whether natural classes exist or not (Quine 1970; Dennett 1991). When searching for the foundations of categorization, we assume no distinction between natural and artificial or other implicit categorizations (Du Châtelet 1759). Instead, we reason that everything can be categorized as we adopt the old atomistic tenet where everything is ultimately composed of basic building blocks, the building block we identify to be the quantum of actions (Pernu and Annila 2012; Annila 2010; Annila 2012; Annila 2016b). Our hypothesis is axiomatic and falsifiable. It can be proven wrong by showing that there is, in fact, an entity which cannot be broken down to the quanta of actions.

The quantum of light, i.e., the photon, is a familiar example of the quantum. Its attributes energy E and period t combine in an invariant measure known as Planck's constant:

$$h = Et. \tag{1}$$

The fixed quantity means that the photon is an indivisible entity. We only assume that everything that exists comprises the quanta. Thus, a compound entity, from now on referred to as a system, integrates its n constituent quanta into an invariant known as the action (Noether 1918)

$$nh = \int 2K dt \tag{2}$$

over the paths of kinetic energy $2K$. This implies to us that conserved quantities in multiples of h are natural categories. For example, the hydrogen atom in its ground state, defined by Eq. (2), is an action with a fixed number of quanta. When the atom absorbs one quantum of light, it will change from the ground state category to an excited state category. We acknowledge that in practice, it is not easy to keep track of all quanta in a given system in its universal surroundings (Annila 2016b; Lewis 1926; Annila 2015; Abbott et al. 2016; Grahn et al. 2018).

It follows from the atomistic axiom that we may formally describe any system in terms of quanta. We do this formally by placing the system's entities as compound quantized actions on levels of an energy diagram. In this scale-free manner, each entity can be assigned to a class by its energy attribute (Du Châtelet 1759). Energetically indistinguishable entities populate the same level in the diagram (Fig. 1). Accordingly, microstates, i.e., permutations of these identical entities, are energetically equivalent. We include dynamics of the system by flows of quanta from one level to another and from the surroundings to the system and *vice versa*.

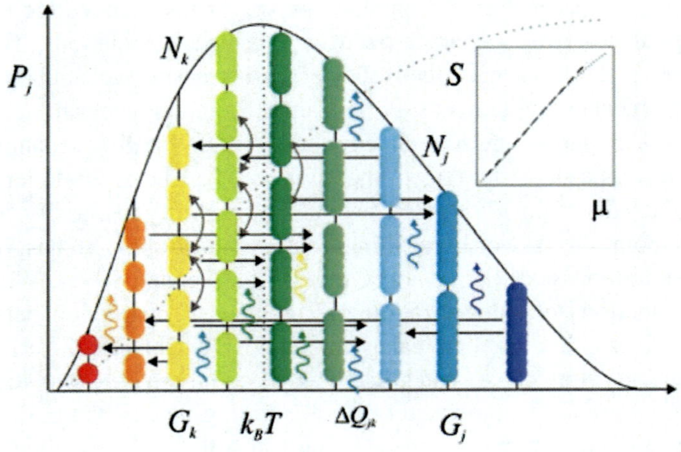

Fig. 1 The system of classes is depicted in terms of an energy level diagram (Koivu-Jolma and Annila 2018). At each level, indexed by k, there is a population, i.e., a class of N_k entities each with energy G_k. The size of N_k is proportional to probability P_k. When an entity in the population N_k transforms into an entity in the population N_j, horizontal arrows indicate paths of available transformations from one class to another. Vertical wavy arrows denote concurrent changes driven by energy in light. The vertical bow arrows mean the exchange of indistinguishable entities without changes in energy, and hence without a change in class. The system evolves, step by step, via absorptive or emissive jk-transformations that are mediated or catalyzed by entities themselves, toward a more probable partition of entities, i.e., a classification, eventually arriving at a stationary-state balance where the levels are populated so that the average energy k_BT equals that in the system's surroundings. A sufficiently statistical system will evolve gradually because a single step of absorption or emission is a small perturbation of the average energy. Hence at each step of evolution, the outlined skewed quasi-stationary partition does not change much. This maximum-entropy distribution accumulates along a sigmoid curve (dotted) which is on a log-log scale (insert) a straight line of entropy S vs. [chemical] potential energy μ

This allows us to describe the evolution of classification along with the system's evolution.

We exemplify our general nomenclature with a molecular system. The energy that is bound in a population [of molecules] is given by chemical potential $\mu_j = k_BT\ln N_j + G_j$ where N_j is the number of entities [molecules], G_j is the energy of one entity, and k_BT is the average energy of the system at temperature T. The categorization of entities in terms of the energy diagram allows us to formulate the state of a system, i.e., to define categories. The change from one category to another involves a change in energy. The classification itself will also cause changes because at least one quantum must be obtained from the entity to quantify its class. It is insightful to speak in thermodynamic terms because then we are free from human categorization.

3 Theory

Classification entails that the classes are in some relation to each other. For example, individuals in a population can be categorized by body weight, in other words, put on the same scale relative to each other. Ultimately any relation can be given in terms of energy (Fig. 1), and hence, there is according to the second law of thermodynamics also an optimal occupancy of various classes. This free energy minimum manifests itself in scale-free patterns, i.e., nearly lognormal distributions, including logarithmic spirals, that accumulate along sigmoid curves and at times display oscillations and even chaotic trajectories (Du Châtelet 1759; Kapteyn 1903; Gaddum 1945; Limpert et al. 2001; Grönholm and Annila 2007). This conclusion about natural classification can be drawn from the probability theory of many-body systems (Sharma and Annila 2007; Du Châtelet 1759).

3.1 Definitions

Let us consider the probability P_j that a class, indexed with j, is populated by N_j entities. For example, we may consider a species with individuals in an ecosystem or a molecular species in a chemical reaction mixture. First P_j depends on energy $\mu_k = k_BT\ln N_k + G_k$ that is bound to the necessary substrates in numbers N_k each with energy G_k. This means, for example, that for predators to exist, there must be preys. Obviously, the general formalism accounts for all species of a food web. This energy transduction network roots from the photon absorption of sunlight and it terminates at dissipation of photons to the cold space. Thus, P_j depends also on the influx or efflux of energy, i.e., dissipation that couples to changes in the population from N_k to N_j and *vice versa*. Namely, the predators cannot consume the preys without some dissipation.

Likewise, a chemical reaction from substrates to products cannot proceed without either emission or absorption of photons. In fact, it follows from our axiom that no change of state can take place without either absorption or emission of at least quantum. The flux of quanta (photons) is denoted by the energy difference $i\Delta Q_{jk}$, which matches the energy difference $\Delta G_{jk} = G_j - G_k$ per entity between the k-substrates and j-products. The imaginary part merely indicates that the vector potential from the surroundings to the system or *vice versa* is orthogonal to the scalar [chemical] potential.

The probability P_j for the population N_j

$$P_j = \left[\prod_{k=1} N_k \mathrm{e}^{-\Delta G_{jk}/k_BT} \mathrm{e}^{+i\Delta Q_{jk}/k_BT} \right]^{N_j} / N_j! \tag{3}$$

is obtained as the product of its k-substrates including an influx of photons that couple to the jk-transformations. For example, any chemical reaction is either endo- or exoenergetic. Thus, metabolism of a predator can be described accordingly. In Eq. (3) the division by factorial $N_j!$ enumerates the inconsequential exchange of identical entities that causes no changes in the classification scheme (Fig. 1). If any one vital k-ingredient is missing altogether from the product form Π_k, the j-population cannot exist, i.e., $P_j = 0$. Similarly, if no flux of energy couples from the surroundings to the system, the jk-transformation cannot take place. This means, for example, that when a vital nutrient is missing altogether, the species cannot proliferate even if everything else would be available. The index includes transformation stoichiometry by running from $k = 1$ to an unknown upper limit that is eventually reached when the system has attained thermodynamic balance with its surroundings. In a small chemical mixture, it might be possible to determine the stoichiometry of all conceivable reactions, but many a system is too big and diverse to imagine all possible evolutionary scenarios. For example, it is very difficult to predict how the metabolic system of a bacterium will respond to various agents, for example, intended to limit bacterial infection. Nonetheless Eq. (3) formally keeps track of all ingredients down to the precision of a single quantum.

The probability for the population of any other class can be expressed likewise. Thus, the total probability for the populations in all classes is the product of P_j:s:

$$P = \prod_{j=1} P_j = \prod_{j=1} \left[\prod_{k=1} N_k \mathrm{e}^{-\Delta G_{jk}/k_B T} \mathrm{e}^{+i\Delta Q_{jk}/k_B T} \right]^{N_j} / N_j!. \tag{4}$$

In this manner, the total probability provides an energetic status, e.g., of a cellular system, ecosystem, or economic system. The status is high in energy-rich surroundings. Under such circumstances, the classification yields a high number of species or products distinguished from each other by numerous energy differences. Conversely, in surroundings that are low in energy, the most probable state of a system contains only relatively few entities. In other words, the probability is the system's energetic measure in relation to its surrounding systems, so that the highest value is attained at thermodynamic balance. Thus, natural categories are neither arbitrary nor algorithmic but relate to the surroundings in energetic terms.

The logarithm of P ($\ln P$), rather than P, is an additive measure to quantify the energetic optimality of a given categorization. Then, one classification can be compared with another by comparing the sums $\Sigma \ln P_j$. For example, a finer decimation of entities in distinct classes will yield a higher value than a coarse one. This means that the categorization is invariably a subjective process in accordance with observations. For historical reasons, entropy S is defined as the logarithm of P

$$S = k_B \ln P = k_B \ln\left[\prod_{j=1}\left(\prod_{k=1} N_k e^{-\Delta G_{jk}/k_B T} e^{+i\Delta Q_{jk}/k_B T}\right)^{N_j} / N_j!\right]$$
$$= \frac{1}{T}\left[\sum_{j=1} N_j k_B T + N_j\left(\sum_{k=1}\mu_k - \mu_j + i\Delta Q_{jk}\right)\right] \tag{5}$$

when multiplied by Boltzmann's constant k_B. It is the additive measure for natural classification. In Eq. (5) Stirling's approximation $\ln N_j! \approx N_j \ln N_j - N_j$ has been used. The approximation is consistent with the statistical character of a system. Specifically, if there were only a few objects, their categorization would be troublesome, to begin with.

It is worth emphasizing that entropy (Eq. 5), when multiplied with temperature T, identifies classes on the basis of two terms: first by energy $\Sigma_j N_j k_B T$ that is bound in the j-populations of the classes (Kondepudi and Prigogine 1998) indexed with j and second by energy $\Sigma_j N_j(\Sigma_k \mu_k - \mu_j + i\Delta Q_{jk})$ that still is present between the system and its surroundings. The first term $\Sigma_j N_j k_B$ is the familiar entropy obtained from statistical mechanics for a closed or stationary system. Obviously, when all energy is bound in the various populations, the classes are steady and thus unambiguously countable. At this maximum entropy state, there is no net flow of energy carriers between the system and its surroundings, and hence neither a new class will appear, nor an old one will disappear.

Conversely, the second term $\Sigma_j N_j(\Sigma_k \mu_k - \mu_j + i\Delta Q_{jk})/T$ means that the classification system is open for evolution by consuming energy differences relative to its surroundings, i.e., forces that motivate classification. This flux of energy carriers from the system to its surroundings, or *vice versa*, leads to the increase in entropy until all energy differences have leveled off. The free energy term means, for instance, that there is a force that drives further or finer classification. Alternatively, there may not be enough free energy to maintain the current degree of classification, but the classes will be merged to regain balance with resources. This is, of course, common sense. A finer classification needs more resources than a course one.

3.2 *Classification as a Process*

The natural evolution of the classification scheme will be obtained from the differential equation of motion for entropy (Eq. 5):

$$\frac{dS}{dt} = \sum\nolimits_{j=1} \frac{dS}{dN_j}\frac{dN_j}{dt} = \frac{1}{T}\sum\nolimits_{j=1} \frac{dN_j}{dt}\left(\sum\nolimits_{k=1}\mu_k - \mu_j + i\Delta Q_{jk}\right) \geq 0 \tag{6}$$

where the chain rule has been used. The two-term product reveals that the population N_j, of the class j, will change by $d_t N_j$ proportional to the driving force $A_j = \Sigma_k \mu_k -$

$\mu_j + i\Delta Q_{jk}$. Thus, the measure for classification can only increase, i.e., $dS \geq 0$, since $(A_j)^2 \geq 0$. In other words, the classification will evolve as long as there are motive forces for it and means to improve it. From this perspective, the predator is likely to evaluate its prey carefully when it must invest a considerable amount of resources in catching it. This is also common knowledge. The predator will not attack arbitrarily but consider which prey it will try to catch. Conversely, when surrounded by abundant resources, the degree of categorization is expected to be low. Then basically anything will do. For example, many whale species simply swim around with their mouths open and filter food through their baleen bristles, when they have found a rich school of fish or krill.

It is worth emphasizing that the classification will progress to define finer details only when such subtle differences contribute to the overall free energy consumption. Put differently, the finer classification must provide benefits that supersede its costs. Otherwise, it will not be adopted. Conversely, the classification scheme will evolve by abandoning classes when the distinction is energetically unfavorable or even immaterial. For example, many languages, when adapting to the modern way of life, are rapidly losing vocabulary related to the old rural lifestyle. At times the changes in surroundings are so big and rapid that the changes in categorization display oscillations and even chaotic characteristics. For example, words will acquire new distinct meanings among subpopulations, and hence due to decreased communication societal cohesion decreases overall.

Finally, when the classification has consumed all forms of free energy, the class structure has attained thermodynamic balance, i.e., $dS = 0$. The optimal classification has converged in a free energy minimum. It is Lyapunov-stable so that any perturbation δN_j away from a steady-state population N_j^{ss} will cause decrease in $S(\delta N_j) < 0$ and concurrently increase in $d_t S(\delta N_j) > 0$ (Strogatz 2000). In other words, the further away N_j would be from N_j^{ss}, the larger will be the restoring force A_j. This balance manifests itself, for example, in maintaining consensus about meanings of words. The quest for the free energy minimum categorization is customarily understood so that a useful classification mechanism is such that knowledge accurately infers object properties and these properties accurately infer object classes (Corter and Gluck 1992). Likewise, many species in stable ecosystems have highly specialized diets.

It is worth pointing out that our approach for denoting a hierarchy in thermodynamics terms is not unique (Bar-Yam 2004a; England 2013; England 2015) and not the only option either (Allen et al. 2017). The emergence of new classes has been addressed in mathematical terms (Bar-Yam 2004b).

3.3 *The Subjective Character of Classification*

The natural class structure that extends down to single quanta is obviously inaccessible in practice to a subject. Thus, one's categorization is invariably narrow and coarse grained. It limits to one's own observations and inferences as well as influences obtained from others. In other words, one's categorization is limited by resources and biased by past processes. This behavior is recognized as cognitive and confirmation biases as well as at the level of systems, as systemic or institutional bias (Nickerson 1998; Kahneman and Shane 2002; Anttila and Annila 2011). Nevertheless, the subjective classification is invariably governed by the second law of thermodynamics (Eq. 6). Put differently, the subjective classification, while narrow and coarse, is not arbitrary but energetically optimal for the subject. This revelation prompts us to analyze individual classification schemes for meanings as well as for inconsistencies.

One makes sense of perceptions by categorizing them. According to the second law of thermodynamics, making sense means ultimately consuming free energy (Annila 2016a; Anttila and Annila 2011; Annila and Salthe 2009). Conversely, from nonsense one cannot benefit [energetically]. Only some dissimilarity among observations will prompt categorization. Surely it makes a difference to distinguish an edible plant from a poisonous one. According to the thermodynamic tenet free energy motivates one to make distinctions of any kind. Conversely, when the reward for one in categorization is minimal, it will not take place. Approval for this stance is often sought by asking, "Who cares?"

It is intriguing that a subject may insist on making a difference among objects when there is no solid ground for it. For example, one tends to partition nature to animate and inanimate, although there is no single attribute that would warrant such a distinction. This is to say that many an illusory classification is motivated by quantitative rather than qualitative differences. The deceptive division is practical, but it leads to an inconsistent worldview. In terms of physics, inconsistency in classification is a tension, i.e., a force that finds no way to break out. Thus, the puzzle about "What is life?" prevails as long as one insists on having distinct classes for a living and nonliving against all evidence. Although science has abandoned vitalism eons ago (Wöhler 1828; Annila and Baverstock 2014; Annila and Kolehmainen 2015), this kind of fundamental questions are still deemed as philosophical. We wish to point out that they are, in fact, physical when everything is considered as being composed of quanta.

Also, the curious case when there is a difference, but the subject fails to make one, is also worth clarifying. For example, it is quite common that one fails to distinguish two somewhat similar sounds in a foreign language when the two are not distinct and present in one's native tongue. It takes extra effort to learn to hear the difference. Likewise, many other things are often placed in preexisting categories by presumptions and resemblances rather than putting effort into refining one's categorization. Thus, one easily loses opportunities to benefit from making the distinction between superficially similar perceptions. This also demonstrates that classification is a process governed by the energetic imperative of maximal efficiency, but efficiency is rated by the subject who performs the classification.

Actual disputes about definitions and meanings, i.e., differences in classification, are quite common among people. Although it may not be so obvious, the objective of a quarrel is to work out a common scheme of classification, i.e., an agreement on how to rationalize the state of affairs. In terms of physics, common categories allow a coherent and integrated consumption of free energy. First, when the optimal path along the resultant force has been agreed upon, it can be pursued. Of course, the agreement is motivated only when the gain in free energy consumption can be seen to exceed the energetic costs involved in the common category-making. In modern societies, these expenses are typically the costs of standardization (Annila and Salthe 2009; Annila and Salthe 2010). Therefore, those ones with least class structure are most apt to adopt a new classification whereas those with the already well-established classification scheme will find it unrewarding to invest in a new way of thinking or doing. By these examples, we wish to point out that physics has a say in social sciences as well when formulated for open, evolving systems (Koivu-Jolma and Annila 2018; Anttila and Annila 2011; Annila and Salthe 2009; Annila and Salthe 2010).

Finally, it is of interest to note that since Eq. (6) also describes oscillations and even chaotic trajectories, these characteristics are expected to manifest themselves also in categorization (Sornette 2006). The oscillations in categorization are, in fact, quite common. For example, many words in English will be categorized as either verbs or nouns, depending on the context. In general terms of physics, the context is the surroundings that ultimately dictate the meaningful classification, i.e., least-time free energy consumption.

Chaos in categorization is expected when the surroundings vary widely. When "rules" are repeatedly changing, it will be hard to root one scheme of categorization over and others. In other words, category-making fails. The chaotic behavior in categorization can be modeled by the logistic map (Eidenberger 2014). In turn, it has been shown to approximate the least-time free energy consumption. Chaos is typical when a whole class structure collapses. When relationships between classes are obscure, acts will be arbitrary.

We realize that our derivation of natural classes and classification from the principle of physics may, at first sight, appear somewhat remote to contemporary theories of categorization. Therefore, we will work out the correspondence with the most common tenets of classification.

4 Discussion

4.1 Correspondence with Conceptual Classification

Aristotle's categorizing by narrowing questions successively can be put in an algorithmic form, known as conceptual clustering (Michalski and Stepp 1983; Carpineto and Romano 1993; Fisher and Pazzani 1991). The clustering algorithm predefines the path of categorization. In this sense, the algorithm mimics the evolutionary path toward the optimal categorization as given by Eq. (6). However, the conceptual clustering is a deterministic model, whereas the categorization process is non-determinate because the category-making itself affects the categories and *vice versa* (Biswas et al. 1998). Put differently, it is not possible to know in advance what will be encountered and how the encounters will, in turn, affect further encounters. Mathematically speaking variables cannot be separated in the evolutionary equation (Eq. 6), and hence, it cannot be solved (Du Châtelet 1759). We expect machine learning to benefit from this profound insight (Eidenberger 2014). In fact, machines are already thought by exposing them to large amounts of data, i.e., with experience, rather than by programming them to encounter conceivable situations.

Despite its disadvantage in complying with non-determinate reality, the algorithmic approach will suit many a purpose of categorization by being a handy model, i.e., computable in polynomial time. A perhaps more troublesome shortcoming of the algorithmic classification is the lack of energetically defined target function, i.e., the least-time free energy consumption. Then the class structure may evolve in a nonnatural way, for instance, by combining letters to words with no meaning. We expect the algorithmic classification to fail when meanings disperse widely. For example, when symptoms are diagnosed, it is not only subtle differences in observations and laboratory tests that matter but also the consequences of classification. Namely, when the categorization misses a fatal but rare disease, the difference in the data might well be insignificant, but the difference in the outcome is dramatic.

Surely, this problem of meaning in classification has been recognized. The quest for the free energy minimum has been modeled by assigning each class with utility whose maximization drives the clustering formation (Gennari et al. 1989; Lebowitz 1987). Thus, the utility maximization mimics entropy maximization, in fact also by its functional form when given by Kullback–Leibler divergence (Kullback and Leibler 1951). Nevertheless, the model's probability for two objects to be in the same or different category is not expressly given in energetic terms as in Eq. (3), but by phenomenological attributes. Also, it is worth stressing that the category utility sets in advance a deterministic layout. Thus, the method is biased, but its effectiveness is of great practical value.

The conceptual clustering as a classification method is closely related to data clustering. The probabilistic COBWEB algorithm (Fisher 1987; Fisher and Langley 1986) organizes observations into a classification tree. Each tree node represents a class and is summarized by a probabilistic attribute-value distribution under the node. This mode of organization corresponds qualitatively to the energy level diagram (Fig. 1), which can also be presented as trees and networks. On the one hand, the open structure allows one to describe any concept as well as to predict missing objects or to classify new objects (Iba and Langley 2011). On the other hand, there is no unambiguous principle to choose parameters of the algorithmic categorization that may even end up with classification produced by binary yes/no classification (Talavera and Béjar 2001). For example, an entity, say, a bacterial species, is recognized as a member of the class, i.e., a specific taxon, when the sum of predefined class attributes exceeds a given threshold. Obviously, it takes prior knowledge about the diversity of attributes to set a meaningful threshold. Moreover, meaningfulness itself will depend, for instance, on consequences of misclassification, for example, on a wrong diagnosis following from a mistyped bacterial strain. The problem of setting the thresholds is particularly pronounced in automated classification when the machine has no sense of meanings, i.e., consumption of free energy (Karnani et al. 2009). The poor sense of meanings does not limit to the machines.

4.2 Correspondence with Prototype Theory

Prototype theory is a type of graded categorization, which groups identities based on prototypes (Osherson and Smith 1982; Lakoff 1989; Lakoff 1987). A prototype (Rosch 1983; Smith and Minda 2002) is defined as a stimulus that takes a salient position in a class, later redefined as the most central member of a class. Prototype theory is a step away from definition-based models. For example, prototype theory would consider a class like an atom consisting of different entities each with unequal status, e.g., a hydrogen atom is more *prototypical* of an atom than say a niobium atom. This approach is cognitive in the sense that it accepts that categories are graded and inconsistent, but as we argue, ultimately commensurable in energetic terms. The prototype theory can describe even abstract classes, but by our naturalistic tenet, everything is ultimately embodied by quantized actions. The inherent subjectivity of the approach can be exemplified by categories that are different for separate cultures (Smith et al. 1988).

Clearly, also the prototype theory parallels our thermodynamic theory of classes and classification. The most central member of a class is a natural notion for distribution whose central value is given by the average energy (k_BT). Moreover, the subjective character is also inherent in the natural classification.

The prototype theory can also be described in terms of dynamic systems theory where a given object is assigned with a weight determined by past conditions and depending on current conditions (Langacker 1987). Thus, a category reflects how it has been employed in the past. This way prototype systems allow for changes in meaning which are common to languages (Wittgenstein 1958). This path dependence parallels our natural classification.

The recursive nature of prototype systems resembles mathematical iteration. Consequently, outcomes inflate over time, and hence also category definitions keep changing. In other words, prototype systems are nonlinear due to feedback mechanisms. The nonlinearity, e.g., in Eq. (6), is also a characteristic of the natural classification. Expressively a small cause can produce a substantial, even a chaotic, effect.

5 Conclusion

Categorization is such an innate faculty of human beings that one hardly pays attention to it. In fact, the ability to distinguish one from another, as well as to group entities that are alike, appears to be vital for our survival. This evolutionary perspective implies that also other species behave similarly, and hence, we reason that the category-making is not distinctive to humans. Here we have extended this conclusion further by using the thermodynamic theory of evolution that there is an ultimate definition of a class by the quantum of action, which is the basic building block of nature. Moreover, we conclude that there is an optimal way to place objects and observations in classes. This imperative is known as the second law of thermodynamics. Thus, we understand categorization to equate ultimately with least-time free energy consumption, which is known in biological terms as survival.

Our comprehension of the ultimate classes and optimality of classification is convergent with observations that modern cultures aim for an even better understanding of the world by proceeding toward ever-finer decimation and by building ever-larger hierarchical systems and doing it ever faster. This holistic tenet provides an eye-opening viewpoint to human activities by revealing that they are after all not unique to humans and animates either.

References

Abbott BP et al (2016) (LIGO Scientific Collaboration and Virgo Collaboration) observation of gravitational waves from a binary black hole merger. Phys Rev Lett 116:131103. https://doi.org/10.1103/PhysRevLett.116.061102

Allen B, Stacey BC, Bar-Yam Y (2017) Multiscale information theory and the marginal utility of information. Entropy 19:273–311. https://doi.org/10.3390/e19060273

Annila A (2010) All in action. Entropy 12:2333–2358. https://doi.org/10.3390/e12112333

Annila A (2012) The meaning of mass. Int J Theor Math Phys 2:67–78. https://doi.org/10.5923/j.ijtmp.20120204.03

Annila A (2015) The substance of gravity. Physics Essays 28:208–218. https://doi.org/10.4006/0836-1398-28.2.208

Annila A (2016a) On the character of consciousness. Front Syst Neurosci 10:27. https://doi.org/10.3389/fnsys.2016.00027

Annila A (2016b) Natural thermodynamics. Physica A 444:843–852. https://doi.org/10.1016/j.physa.2015.10.105

Annila A, Baverstock K (2014) Genes without prominence: a reappraisal of the foundations of biology. J Roc Soc Interface 11:20131017. https://doi.org/10.1098/rsif.2013.1017

Annila A, Kolehmainen E (2015) On the divide between animate and inanimate. J Sys Chem 6:1–3. https://doi.org/10.1186/s13322-015-0008-8

Annila A, Salthe S (2009) Economies evolve by energy dispersal. Entropy 11:606–633. https://doi.org/10.3390/e11040606

Annila A, Salthe S (2010) Cultural naturalism. Entropy 12:1325–2343. https://doi.org/10.3390/e12061325

Anttila J, Annila A (2011) Natural games. Phys Lett A 375:3755–3761. https://doi.org/10.1016/j.physleta.2011.08.056

Bar-Yam Y (2004a) Multiscale complexity/entropy. Advs Complex Syst 7:47–63. https://doi.org/10.1142/S0219525904000068

Bar-Yam Y (2004b) A mathematical theory of strong emergence using multiscale variety. Complexity 9:15–24. https://doi.org/10.1002/cplx.20029

Berryman S (2008) Ancient atomism. In: Zalta EN (ed) The Stanford encyclopedia of philosophy (Fall 2008 Edition). http://platostanfordedu/archives/fall2008/entries/atomism-ancient/

Biswas G, Weingberg JB, Fisher DH (1998) Iterate: a conceptual clustering algorithm for data mining. IEEE Trans Syst Man Cybernetics Part C Appl Rev 28:100–111

Carpineto C, Romano G (1993) GALOIS: an order-theoretic approach to conceptual clustering. Proc ICML:33–40

Chalmers A (2014) Atomism from the 17th to the 20th century. In: Zalta EN (ed) The Stanford encyclopedia of philosophy (Winter 2014 Edition). http://platostanfordedu/archives/win2014/entries/atomism-modern/

Corter JE, Gluck MA (1992) Explaining basic categories: feature predictability and information. Psychol Bull 111:291–303

De Maupertuis P-LM (1746) Les loix du mouvement et du repos déduites d'un Principe metaphysique. Histoire de l'Académie Royale des Sciences et des Belles-Lettres de Berlin 1746:267–294

Dennett DC (1991) Real patterns. J Philos 88:27–51. https://doi.org/10.2307/2027085

Dretske FI (1977) Laws of nature. Philos Sci 44:248–268

Du Châtelet E (1759) Institutions de physique. (Prault, Paris France 1740) Facsimile of 1759 edition: Principies mathématiques de la philosophie naturelle. I–II Éditions Jacques Gabay, Paris, France

Eidenberger H (2014) Categorization and machine learning: the modeling of human understanding in computers. Books on Demand, Germany

England JL (2013) Statistical physics of self-replication. J Chem Phys 139:121923. https://doi.org/10.1063/1.4818538

England JL (2015) Dissipative adaptation in driven self-assembly. Nat Nanotech 10:919–923. https://doi.org/10.1038/nnano.2015.250

Fisher DH (1987) Knowledge acquisition via incremental conceptual clustering. Mach Learn 2:139–172. https://doi.org/10.1023/A:1022852608280

Fisher DH, Langley PW (1986) Conceptual clustering and its relation to numerical taxonomy. In: Gale WA (ed) Artificial intelligence and statistics. Addison Wesley, Reading, MA, USA

Fisher DH, Pazzani MJ (1991) Computational models of concept learning. In: Fisher DH, Pazzani MJ, Langley P (eds) Concept formation: knowledge and experience in unsupervised learning. Morgan Kaufmann, San Mateo, CA, USA, pp 3–43

Gaddum JH (1945) Lognormal distributions. Nature 156:463–466. https://doi.org/10.1038/156463a0

Gennari JH, Langley PW, Fisher DH (1989) Models of incremental concept formation. Artif Intell 40:11–61. https://doi.org/10.1016/0004-3702(89)90046-5

Grahn P, Annila A, Kolehmainen E (2018) On the carrier of inertia. AIP Adv 8:035028. https://doi.org/10.1063/1.5020240

Grönholm T, Annila A (2007) Natural distribution. Math Biosci 210:659–667. https://doi.org/10.1016/j.mbs.2007.07.004

Heisenberg W (1927) Über den anschaulichen Inhalt der quantentheoretischen Kinematik und Mechanik. Z Phys 43:172–198. https://doi.org/10.1007/BF01397280

Iba W, Langley P (2011) Cobweb models of categorization and probabilistic concept formation. In: Pothos EM, Willis AJ (eds) Formal approaches in categorization. Cambridge University Press, Cambridge, MA, USA

Jackson P (1998) Introduction to expert systems. Addison Wesley, Harlow, UK

Jordan P (1934) Quantenphysikalische bemerkungen zur biologie und psychologie. Erkenntnis 4:215–252

Kahneman D, Shane F (2002) Representativeness revisited: attribute substitution in intuitive judgment. In: Gilovich T, Griffin D, Kahneman D (eds) Heuristics and biases: the psychology of intuitive judgment. Cambridge University Press, Cambridge, MA, USA

Kapteyn JC (1903) Skew frequency curves in biology and statistics. Astronomical Laboratory, Noordhoff, Groningen, The Netherlands

Karnani M, Pääkkönen K, Annila A (2009) The physical character of information. Proc R Soc A 465:2155–2175. https://doi.org/10.1098/rspa.2009.0063

Koivu-Jolma M, Annila A (2018) Epidemic as a natural process. Math Biosci 299:97–102. https://doi.org/10.3390/e12061325

Kondepudi D, Prigogine I (1998) Modern thermodynamics: from heat engines to dissipative structures. John Wiley & Sons Ltd, Chichester, UK

Kullback S, Leibler RA (1951) On information and sufficiency. Ann Math Stat 22:79–86. https://doi.org/10.1214/aoms/1177729694

Lakoff G (1987) Women fire and dangerous things: what categories reveal about the mind. The University of Chicago Press, Chicago IL, USA. https://doi.org/10.7208/chicago/9780226471013.001.0001

Lakoff G (1989) Cognitive models and prototype theory. In: Margolis E, Laurence S (eds) Concepts: Core readings. MIT Press, Cambridge, MA, USA

Langacker R (1987) Foundations of cognitive grammar volume 1: theoretical prerequisites. Stanford University Press, Stanford, CA, USA

Lebowitz M (1987) Experiments with incremental concept formation. Mach Learn 2:103–138. https://doi.org/10.1023/A:1022800624210

Lewis GN (1926) The conservation of photons. Nature 118:874–875. https://doi.org/10.1038/118874a0

Limpert E, Stahel WA, Abbt M (2001) Log-normal distributions across the sciences: keys and clues. Bioscience 51:341–352. https://doi.org/10.1641/0006-3568(2001)051[0341:LNDATS]2.0.CO;2

Mäkelä T, Annila A (2010) Natural patterns of energy dispersal. Phys Life Rev 7:477–498. https://doi.org/10.1016/j.plrev.2010.10.001

Michalski RS, Stepp RE (1983) In: Michalski RS et al (eds) Learning from observation: conceptual clustering in machine learning: an artificial intelligence approach. TIOGA Publishing Co, Palo Alto, CA, USA

Nickerson RS (1998) Confirmation bias: a ubiquitous phenomenon in many guises. Rev Gen Psychol 2:175–220

Noether E (1918) Invariante Variationsprobleme Nachrichten von der Gesellschaft der Wissenschaften zu Göttingen. Math-Phys Kl 1918:235–257.

Osherson DN, Smith EE (1982) On the adequacy of prototype theory as a theory of concepts. Cognition 9:35–58. https://doi.org/10.1016/0010-0277(81)90013-5

Palmer J (2012) Parmenides. In: Zalta EN (ed) The Stanford encyclopedia of philosophy. http://platostanfordedu/entries/parmenides/

Pernu TK, Annila A (2012) Natural emergence. Complexity 17:44–47. https://doi.org/10.1002/cplx.21388

Quine WV (1970) Natural kinds. In: Rescher N et al (eds) Essays in honor of Carl G. Hempel. D. Reidel, Dordrecht, The Netherlands, pp 41–56

Rosch E (1983) Prototype classification and logical classification: the two system in new trends in conceptual representation. In: Scholnick EK (ed) Challenges to Piaget's theory? Lawrence Erlbaum, Hillsdale, NJ, USA, pp 73–86

Salthe SN (1993) Development and evolution: complexity and change in biology. MIT Press, Cambridge, MA, USA

Sharma V, Annila A (2007) Natural process – natural selection. Biophys Chem 127:123–128. https://doi.org/10.1016/j.bpc.2007.01.005

Smith JD, Minda JP (2002) Distinguishing prototype-based and exemplar-based processes in dot-pattern category learning. J Exp Psychol Learn Mem Cogn 28:1433–1458. https://doi.org/10.1037/0278-7393.28.4.800

Smith EE, Osherson DN, Rips LJ, Keane M (1988) Combining prototypes: a selective modification model. Cogn Sci 12:485–527. https://doi.org/10.1207/s15516709cog1204_1

Sornette D (2006) Critical phenomena in natural sciences: chaos fractals self-organization and disorder: concepts and tools (Springer series in synergetics). Springer, Berlin, Germany

Strogatz SH (2000) Nonlinear dynamics and chaos with applications to physics biology chemistry and engineering. Westview, Cambridge, MA, USA

Talavera L, Béjar J (2001) Generality-based conceptual clustering with probabilistic concepts. IEEE Trans Pattern Anal Mach Intelligence 23:196–206

Ulanowicz RE, Hannon BM (1987) Life and the production of entropy. Proc R Soc Lond B 232:181–192. https://doi.org/10.1098/rspb.1987.0067

Wittgenstein L (1958) Philosophical investigations. Blackwell Publishers, Oxford, UK

Wöhler W (1828) Ueber künstliche Bildung des Harnstoffs. Ann Phys Chem 88:253–256. https://doi.org/10.1002/andp.18280880206

The Operator Theory: A Yardstick for Complexity from Quarks to Memons—Relationships with Evolution and Thermodynamics

G. A. J. M. Jagers op Akkerhuis

Abstract Thermodynamic theory predicts that the universe develops towards maximum energy dispersal. Meanwhile, complex systems continue to form. The search for an explanation of these seemingly opposing trends has inspired many scientists. The theory of nonequilibrium thermodynamics brought much progress, allowing subsystems to become more complex at the costs of external energy gradients. But energy gradients may not tell the whole story, because, while they explain the existence of cells, gradients alone cannot explain the existence of complex organisms such as plants, tigers, or humans. Contributing to our understanding of the relationships between complexity and thermodynamics, this study focuses on a hierarchical subset of all complex systems. The systems in this subset have formed, in a step-by-step way, through a series of "dual-closure" processes. Every system produced through dual closure is called an "operator," and their stringent complexity hierarchy is called the "operator hierarchy." It is demonstrated that the operators can be grouped into three major classes with fundamentally different thermodynamics: (1) abiotic operators resulting from condensation reactions, (2) organisms resulting from contained autocatalysis and competition, and (3) neural network organisms driven by autocatalysis, learning, and competition. To these three groups a fourth group of rapidly evolving systems that are not operators can be added: "artifacts" made by organisms, notably humans. While normally being viewed as the result of self-organization, the design of artifacts may in fact be the product of "allo-organization."

Keywords Operatorhierarchy · O-theory · Thermodynamics · Big evolution · System science · Hierarchy theory

---

G. A. J. M. Jagers op Akkerhuis (✉)
Wageningen University and Research, Wageningen, The Netherlands

G. Y. Georgiev, M. Shokrollahi-Far (eds.), *Efficiency in Complex Systems*,
Springer Proceedings in Complexity, https://doi.org/10.1007/978-3-030-69288-9_3

1 Introduction

How did all the complex systems in the universe come about? In its first moments, the universe was filled with fundamental particles (those of the standard model). Through expansion the universe cooled down. In a cool universe, gravity and condensation reactions allowed early complex structures to take shape, such as hadrons, atoms, molecules, stars, black holes, and planets. Later, more complex structures formed, such as organisms. As a general tendency, one can observe that the complexity of at least *part* of the systems that are newly formed increases over time. The idea behind this study is that a stringent hierarchical ranking of systems according to complexity can offer more insight in the level dependencies of causal processes and thermodynamics.

For investigating the above level dependencies a methodology is needed ranking systems according to "complexity." An independent way of working would be to develop an *a priori* approach that ranks different organizations according to complexity. As a fundamental solution to this ranking challenge, this study uses the operator theory, or OT (Jagers op Akkerhuis and van Straalen 1999). Before discussing how the OT can be used as a yardstick for complexity, I highlight some current insights about complexity, thermodynamic measures, hierarchy, and causality that together sketch the general context for this study.

1.1 Relating Complexity to Thermodynamic Measures

Several studies discuss thermodynamic measures and their relationship with complexity. For example Chaisson (2001, 2011) shows that when systems are ranked according to the moment they first appeared during big history, which ranking is viewed as a measure for complexity, the ranking correlates with a higher amount of free energy degradation, per second, per kilogram, which Chaisson refers to as energy rate density, or ERD (Fig. 1). A related approach is that of Gladyshev (1978), who in the context of "hierarchical thermodynamics" focuses on the reduction per volume of Gibbs free energy (Gibbs 1873), or on the increase in the "stability" of systems (Gladyshev 2017). The fundamental principle of least action (De Maupertuis 1744; Noether 1918) also offers directions for changes in nature (Annila 2010b; Pernu and Annila 2012). Action is the integral over time of the difference between potential and kinetic energy. The idea is that the dynamics of a system self-organize towards least action. The metric of "action efficiency," which quantifies the average action in a system per event, and thus per unit of flow, has been suggested as a measure for self-organization (Georgiev and Georgiev 2002; Georgiev et al. 2015; Georgiev and Chatterjee 2016). Action efficiency is also interpreted as an indication of how "evolved" a system is.

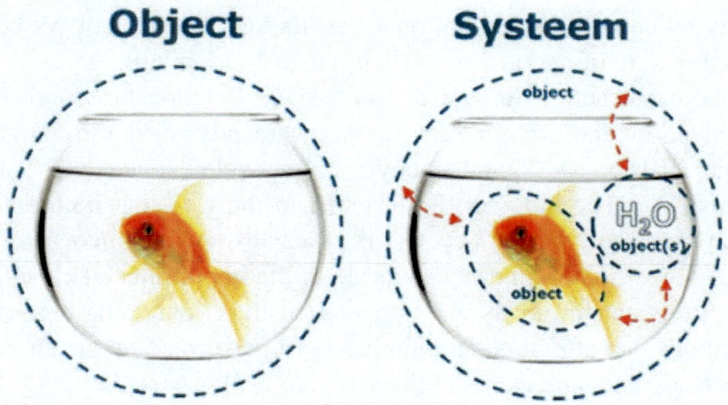

Fig. 1 A bowl with goldfish can be perceived as an object or as a system. The difference results from the *intention* of the observer to view the ensemble either as an object or as a system. Dashed circles indicate objects. Dashed lines indicate relationships (after Jagers op Akkerhuis 2018, 2019)

1.2 Relating Complexity to Hierarchy

To relate complexity to causes and/or thermodynamics, one must start with a theory for hierarchical complexity. Since long, scientists and philosophers have proposed grand theories that rank systems according to "complexity," including Aristotle (384–322 BC, the allegory of the Scala Naturae), Feibleman (1954), Teilhard de Chardin (1959), Laszlo (1972), Ebeling and Feistel (1992), Alvarez de Lorenzana (1993), Chaisson (2001), Pagels (1985) and Spier (1996), Big History), and Maynard Smith and Szathmáry (1995), the major evolutionary transitions).

If the goal is to rank systems according to complexity, one needs criteria for distinguishing entities at one level of complexity from those at other levels of complexity. Knowledge about such criteria has gradually increased throughout history.

The *Scala Naturae* of Aristotle provided a classification, but hardly any information about how entities at different levels had formed. Teilhard de Chardin made the focus more specific, by introducing the idea of "inward complexification," which defines complexity as a product of relationships of elements of a system amongst themselves, in combination with the idea of "centricity," which refers to systems that have a specific "centralized" internal organization. From this basis, he was able to distinguish what he called "true natural units," e.g., the atom, the molecule, the cell, and living beings, from "accidental pseudo units," e.g., a drop of water, a heap of sand, the earth, and the sun (Teilhard de Chardin 1969, p. 137). A few years later Feibleman (1954) suggested a "hierarchy of integrative levels," advocating that all things in such a hierarchy must have physical properties, which statement limits the hierarchy to entities of a material "kind." While focusing on physical entities, he suggests that electrons, protons, and neutrons integrate to atoms, and that atoms integrate to molecules, molecules to cells, and so on. He also suggests: "It would be theoretically possible to assign numbers to the levels to indicate their degree of complexity by counting emergent qualities, though this has not yet been done. As yet we are not sure of the qualities" (Feibleman 1954, p. 60).

Ideas about true natural units with "centralized" relationships, and about qualities that would allow the counting of levels, were advanced through the work on autopoiesis (Varela 1979) and catalytic hypercycle (Eigen and Schuster 1979; Kauffman 1986). The moment that it came to the fore that both autopoiesis and hypercycle represent "closed" cycles of processes, developments started focusing increasingly on closure. An early advocate of the importance of closure, Heylighen (1990) explains closure in its most general/abstract sense: the mathematical term closure implies that after an operation has been performed on the elements of a set, the products are again elements of the set.

Step by step, the concept of closure was elaborated, and various new, less abstract kinds of closure were identified, relating to physical systems. An important distinction was that between function and structure. This distinction plays an important role in Turchin's (1977) work on metasystem transitions (or MST). About this theory Heylighen et al. (1995) states: "Turchin's description (1977, this issue) of an MST contains both a structural aspect, like in Simon's model, and a functional aspect like in Powers's and Campbell's models." Heylighen continues: "The *structural* definition sees a metasystem S' as an integration of a number of subsystems S_i . . . " and "In Turchin's *functional* description, an MST takes place when the activity at the highest control level of some system S becomes itself controlled, forming a higher order system S': control of S = S." In line with this, and focusing on complexity (Heylighen et al. 1995). Heylighen (1999) describes functional complexity as "the complexity produced by differentiation and integration in the *temporal* dimension," and structural complexity as the product of differentiation and integration in the *spatial* dimension. In the website of the Principia Cybernetica project (http://pespmc1.vub.ac.be/WFISSUE.html), and in Heylighen et al. (1995), the term metasystem is introduced. Metasystem transitions function as a "quantum of evolution," a discrete jump to a higher level of complexity. Such quanta provide a principle for ranking evolutionary "progress" or development.

The above developments gave strong impulses to research into closed aspects of the organization of systems. As a rule, however, the results were expressed in the terminology of general systems approaches. This habit dates back to the influential work of von Bertalanffy (1950, 1968) on "general systems theory." It is problematic, however, that approaches in which "systems" produce "systems" or in which interactions produce "higher levels of complexity" are rarely specify the *kinds* of systems, or the *kinds* of complexity involved. For example, when systems produce systems, this holds equally well for atomic systems producing a molecular system, cars producing a traffic system, or humans producing a social system. If one aims at constructing a complexity ranking that selectively includes systems of a special kind, however, one needs an approach which assures that systems of a known major kind will strictly and only produce more complex systems of the same major kind. With this goal in mind, new interpretations of structural and functional closure were developed, and integrated into "dual closure" (Jagers op Akkerhuis and van Straalen 1999; Jagers op Akkerhuis 2008, 2010a, b, 2016). It is because closure offers a highly restrictive combination of criteria for stepwise increases in complexity that it can serve as a foundation for the construction of a stringent complexity hierarchy in which operators produce operators.

1.3 Relating Complexity to Causality

The universe is teeming with complex systems, including celestial bodies, molecules, rivers, organisms, and society. Every increase in complexity of a system, for example from low- to high-complexity molecules, or from organisms with small brains to organisms with large brains, implies more "order" and a reduction of entropy. The reduction of entropy seems to defy the second law of thermodynamics, stating that any process in an overall system leads to an increase in entropy, which, in a material world, corresponds with the dispersal of energy (Annila 2010a, 2010b). In the end, all particles and their energy are dispersed and there is no energy left that can be used to perform work: there is no "free energy" anymore.

As a solution to this puzzle, scientists developed thermodynamic theory about "open" (sub)systems. The term open implies that free energy can be imported as well as exported from the subsystem. Because of such import/export, any open subsystem in the universe can import free energy from energy gradients in the environment, and degrade this energy to perform work. For example plants harvest free energy of the sun, and degrade this to maintain their organization, to create more intricate internal relationships, to grow, and to store energy. Plants and other open subsystems thus exist at the costs of the degradation of free energy gradients: they are special "dissipative" systems (e.g., Prigogine et al. 1972; Prigogine and Stengers 1984; Lambert 1999, 2002, 2007; Lineweaver 2006).

While thermodynamic laws offer limiting conditions and a direction for processes to occur, they neither tell *how* specific physical processes support the formation of any complex system, nor can they predict the *actual form* a complex system will obtain. If such predictions were possible, thermodynamic theory could tell us how the first cells on earth looked like, and how they functioned. Such possibilities would make obsolete all the laborious experiments that currently aim at resolving this long-standing puzzle. As thermodynamic theory does not predict specific organization of complex systems, any analysis of how complex systems have formed must include an analysis of causal processes.

1.4 Outline of this Study

The science of complex systems covers a broad field. Most of the studies in this field deal with large systems consisting of many interacting elements. Examples of major classical topics of inquiry are self-organized criticality (Bak and Sneppen 1993; Bak 1996), relativity, the Big Bang and cosmic inflation (Einstein 1916; Hubble

1929; Penzias and Wilson 1965), the constructal law (Bejan 1997, 2016), chaos and fractals (Mandelbrot 1967; Lorenz 1963; Feigenbaum 1978), tipping points (Holling 1973; Scheffer 2009), power laws in networks (Pareto and Zipf (in Newman 2005; Mandelbrot 1967), Moore's law (Moore 1965), and Darwinian evolution (Darwin 1859; Darwin and Wallace 1858).

In studies such as the above, it is generally the case that complexity is the result of interactions between "agents," "building blocks," "individuals," or "unit systems" of one particular kind, for example a flock of birds, a school of fish, or a society of humans. As long as all the interacting unit systems in a system are of the same kind, and can easily be recognized, there is little use for theory about their complexity. From a broader theoretic perspective, however, one must have an answer to the question: How can one know whether or not specific unit systems are of the same kind? And for systems that are not of the same kind, it is relevant to understand the differences between their complexities, and/or whether, and how, such differences can be ranked hierarchically. The ability of classifying systems according to their "complexity" becomes especially relevant if one aims at answering the question of how complexity is linked to thermodynamics. As was already indicated above, the operator theory, or OT (Jagers op Akkerhuis 2016, Chap. 14.1), focuses specifically on the hierarchy of unit systems. Because the operator theory has such an important position in this study as a basis for furthering our understanding of the relationships between complexity and thermodynamics, I offer a short introduction to this theory in the next paragraphs.

2 The Operator Hierarchy: A Short Introduction

The reasoning in this study leans heavily on the innovative perspective of the operator theory, and how this theory offers causal explanations for the hierarchical formation of special systems called "operators." The intro in the next paragraphs is written from low to high detail. It starts with the term "system," then narrows the focus to the "operators" and their hierarchical construction, and finally explains how the hierarchy of the operators suggests fundamental changes in current classifications of kinds of systems. All this acts as a preparation for later paragraphs about the relationship between complexity and thermodynamics.

2.1 The Term "System"

The concept of a system has many definitions, most of which focus on mental representations of a limited volume of the universe and the interaction between the objects in it. The objects can in turn be viewed as systems consisting of interacting objects, etc. The system and its limits are in principle imaginary, because the distinction between the system and its environment is determined by us. This holds for a car, a cubic centimeter of empty space, as well as an education system. Such considerations inspired Bernard (1865) to conclude that while nature contains physical things, systems exist strictly and only in the minds of people. Taking into account Bernard's perspective on systems, and in line with, e.g., the soft systems approach of Checkland and Scholes (1990), an integrating viewpoint can be constructed suggesting that any analysis starts with a person's *intention* to willfully view something as an object or as a system (Jagers op Akkerhuis 2016). I have named the primitive undecided state, at the moment one can still decide to view a physical/mental entity as a system or as an object, the "sysob" state (Jagers op Akkerhuis 2018). Following an intentional choice for the object viewpoint, a bowl with goldfish represents a single object (Fig. 1, left part). However, if one willfully focuses on internal objects and their interactions, the bowl with goldfish represents a system. When looked at this way, objects and their interactions no longer determine whether a thing "is" a system. Instead, an object becomes a (mental) system as soon as the observer *chooses* to analyze the object *as* a system, by zooming in on the objects in the system, and their interactions.

2.2 *Narrowing the Scope to "Operators"*

Having defined the concept of a system, a next step can be made towards the identification of special kinds of systems, and hierarchical rankings. A property that frequently goes unnoticed, but that is relevant for the majority of classical rankings, is that they rank systems of different logical *kinds*. Some systems are material (e.g., an atom, a molecule, a chair, a car, a person). Other systems are mental groupings of material things (e.g., a population, a society). Still other systems are entirely mental (a unicorn, a deity).

For reasons of logical consistency it is to be preferred that a ranking strictly and only includes systems that all belong to one major kind (Jagers op Akkerhuis 2016, Chap. 16.1, and Jagers op Akkerhuis 2019, Chaps. 1 and 9). For example, in a ranking of material entities, mental entities do not fit in. Any ranking that mixes material and mental entities is no longer consistent in its type. For example, the demand for consistency is not met by the following ranking: atom, molecule, organism, and population. The reason is that the atom and molecule represent material systems, while the term organism is a broad mental class, and the term population is a mental grouping. As the systems in the example are not all of the same major kind, the logic of the ranking is not consistent.

For constructing a consistent complexity ranking, one needs a rule, or a theory, that allows one to select, amidst of all the different kinds of systems, the systems of the desired kind. To deal with this challenge, the operator theory, or OT, makes use of a recursive approach (Jagers op Akkerhuis and van Straalen 1999; Jagers op Akkerhuis 2008, 2012, 2014, 2016, 2019). The recursion starts with small systems of the same kind. Special interactions between the small systems enable the construction of the next larger system. By repeating this story one can identify a sequence of systems of increasingly complex kinds. As all the systems share a common rule for their formation, they can be viewed as belonging to the same major kind. The criteria for any next step in the operator hierarchy are:

1. *Functional closure*. Closure is a mathematical concept that has found its way to system science. I first encountered the term closure in the works of Heylighen (1990) and Kauffman (1993). More recent work on closure in biology can be found in, e.g., Heylighen (1999), Chandler and de Vijver (2000), and Moreno and Mossio (2015). In the OT closure can be either functional or structural. Functional closure implies that operators that are all of the same kind form a cycle of transformation processes. An example of functional closure is the transformations in the atom nucleus where the exchange of pions continuously alters the states of the protons and the neutrons. Another example is the autocatalytic set in a cell, where catalytic molecules transform "resource" molecules into new copies of molecules that together form the autocatalytic set.
2. *Structural closure*. In the OT structural closure describes a layer surrounding the functional closure. One example is the electron shell that surrounds a nuclide, which ensemble is known as an atom. Another example is a membrane around an autocatalytic set, resulting in a cell. The structural closure surrounds the elements of the functional closure and mediates their interactions with the world. As a consequence, these elements can no longer interact freely with the environment.
3. *Obligatory dependency of the functional and structural closure*. At the level of the atoms, this dependency takes the form of the attraction of electrons by the proton, and the shielding off of the atom nucleus from direct interactions with the world by the electron shell. When talking about a cell, the dependency implies that the autocatalytic set produces molecules that sustain the cell membrane, while the cell membrane forms an interface around the molecules of the autocatalytic set, such that they do not diffuse out of the set.
4. *Uniformity*. Uniformity implies that the closures of any next-level operator are based on the closure of the immediately preceding level. No levels in between are accepted. For example the hadrons of the atom nucleus all have the preceding level closure, as do the cells that form a multicellular organism. And a neural network emerges in the cellular environment offered by multicellular closure. There is a single exception to this rule. The electrons of the atom originate from a level below the hadrons. This exception can be viewed as the result of nature not being able to construct an interface of hadrons, and realizing structural closure through the use of particles of the second-next lower closure level.

In fact, the above four criteria can be supplemented with additional criteria for the different kinds of closures that figure in the operator theory. A detailed explanation of this subject leads beyond the goal of the current study. More information on this topic is offered by Jagers op Akkerhuis and van Straalen (1999), and Jagers op Akkerhuis (2010a, 2016).

Of these four criteria, the first two are considered the most paradigmatic. Because of a desire for minimalism, the above criteria are represented by the term "dual closure." In past publications one can observe my terminology of this phenomenon developing from "hypercycle formation" and "compartmentation" (Jagers op Akkerhuis and van Straalen 1999) to "first-next possible closure" (Jagers op Akkerhuis 2010a), to dual closure (Jagers op Akkerhuis 2016).

It may seem now as if the narrow focus on closure would be in the way of a general approach. However, when using a more general approach, one cannot avoid constructing a hierarchy of systems of very different kinds, and complexity levels without clear limits. The OT prevents such problems because a focus on dual closure allows one to identify a construction hierarchy that strictly and only includes operators, while excluding any and all other kinds of systems.

As is explained, e.g., by Jagers op Akkerhuis (2016), if one starts with quarks, successive dual closures allow for the formation of the following series of increasingly complex operators:

- Hadrons
- Atoms
- Molecules
- Cells and multicellulars
- Endosymbiont cells (classically named "eukaryotes"), and multicellulars of endosymbiont cells
- Memons

The OT uses the term "memon" to indicate any organism that in its neural network or in analogous configurations of internal elements stores and processes sensory impressions, concepts, thoughts, or ideas. The name memon refers to the use of the word "memes" in the work of Dawkins (1976) and Blackmore (1999) for stored patterns of learned knowledge, such as words and melodies.

The OT realizes the counting of complexity levels, as imagined by Feibleman (1954): the complexity level of every operator can be counted by focusing on the number of dual-closure steps that preceded it. It is relevant that the stringent hierarchy resulting from dual closure excludes many "levels" that figure in other rankings. It is furthermore relevant that the term operator clearly distinguishes the subset of the operators from all the other systems. The word "operator" is based on the idea that the systems involved operate (as a broad indication for their behavior) in their environment as individual units. The name "operator" is inspired by, but not identical to, the operator concept in the work of Waddington (1969). Clearly, the term operator as used in the OT has a different meaning than in mathematics or in communication.

Based on its dual closure, every operator can be viewed as a single, countable, self-referring system. The term "self" is applicable in a strict sense, because dual closure in several ways defines the operator as a self. The concept of self is discussed in detail by Jagers op Akkerhuis (2019, Chap. 13).

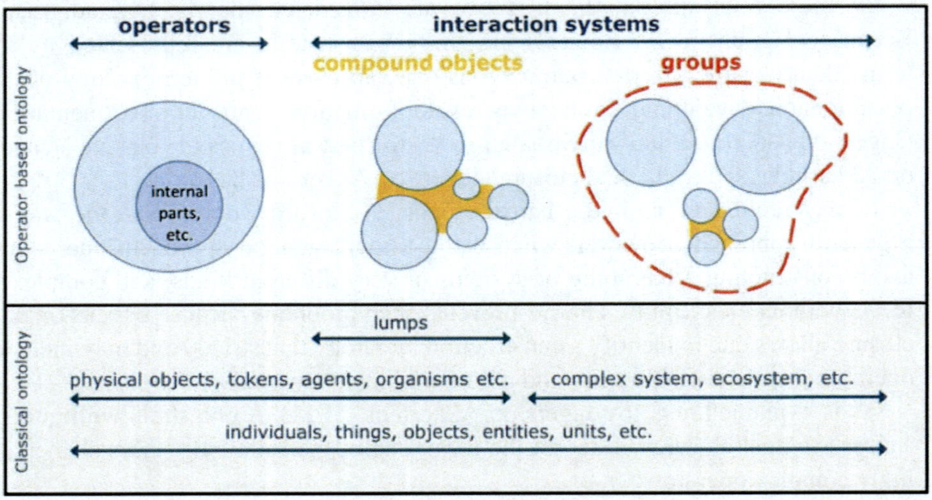

Fig. 2 A top-level ontology for systems/objects. The major classes are "operator" and "interaction system." Interaction systems are subdivided into "compound objects" and "groups." The lower box indicates terms used in classical ontology (modified after Jagers op Akkerhuis 2010a, 2016, 2018, 2019).

2.3 *An Innovative, Stringent Naming of Major Classes of Systems*

The operator theory suggests several fundamental innovations in system science, including a novel classification and naming of the most abstract levels of analysis. The identification and naming of kinds of things in the world can be viewed as a sub-discipline of the philosophy of being, or "ontology." In a narrow sense, the concept of "an ontology" is used for any framework that organizes the terminology people use in a specific scientific discipline (e.g., Vogt et al. 2011; Smith 2004).

The operator theory introduces a novel top-level ontology for system science and complexity research (Fig. 2). The highest level in this ontology comprises all the things that can be perceived of, or can be thought of as a system/object. Next, using the systems perspective, a distinction can be made between systems called *operators* and systems called *interaction systems*. Because of dual closure, every operator can always be counted as a single material unit. In the class of the interaction systems, one finds systems that consist of interacting operators, but the interactions never take the form of dual closure. If an interaction system, or part of it, would develop dual closure, the result would equal the materialization of an operator. The interaction systems are split further into two major classes: *compound objects* and *groups*.

Compound objects are material, and have material parts. These parts can be operators and/or compound objects. The parts of a compound object touch each other physically, while adhering more strongly to each other than to their environment. Examples of *compound objects* are a lump of earth, a piece of wood, a cloth, a stone, a planet and other celestial bodies, a car, a virus particle, and the slug of a slime mold. A freshly dumped heap of stones, for example, does not classify as a compound object, because the stones are not attached more strongly to each other than to their environment (one can lift a single stone).

Finally, the term *group* unites a number of willfully selected, non-adhering, individually countable systems, which can be operators and/or compound objects, that are viewed as belonging together. A group is a *mental* entity. Examples of groups are an ecosystem, a population, a freshly dumped heap of stones, a species, a galaxy, a football team, and a chessboard with pieces.

The top-level ontology of the OT is shown in Fig. 2. This new classification is fundamental to complexity research, because it allows the identification of systems that belong to the same major kind, and in this way prepares the path towards the comparison of within-kind processes, within-kind complexities, and within-kind complexity hierarchies.

The above top-level ontology in Fig. 2 does not include purely mental entities, e.g., a unicorn, or a herd of unicorns. When desired, purely mental entities can be classified as a third major subdivision of all systems. In fact, the classification in Fig. 2 is itself a purely mental entity. This demonstrates that the existence of memons precedes any classification, because classifications and other thoughts develop exclusively in the brains of memons. As they reside in the brain, thoughts cannot interact *directly* with the world. Yet, thoughts play an important *indirect* role in system science, because they affect the behavior of memons, which in turn has many material, "real-world" consequences.

3 Operators of Increasing Complexity: Causes and Thermodynamics

Figure 2 illustrates the distinction between interaction systems on the one hand, and operators on the other. As was indicated in the outline of this study, a large body of past and current complexity research focuses on interaction systems. In the current analyses the focus is on the complexity of systems in the less researched group: the operators. Moreover, the hierarchy of operators can now be used as a unique foundation for studying the relationships between complexity levels and thermodynamics.

To facilitate a structured discussion, the operator hierarchy is split into three parts. It will be shown that the operators of each part differ in major ways in their physical causes and thermodynamics. In addition to these three groups, a fourth group of systems is discussed, that of the "artifacts." Accordingly, the next paragraphs offer an analysis of the relationship between complexity and thermodynamics for three groups of operators, and for artifacts:

1. Operators resulting from condensation
2. Operators based on contained autocatalysis, causing reproduction and competition
3. Operators competing on the basis of neural activity
4. Artifacts resulting from design (and competition)

3.1 Group 1: Operators Resulting from Condensation

Starting with quarks, the formation of operators up to the level of molecules is caused largely by condensation reactions. Basically, condensation reactions are made possible through attractive forces that particles exert on other particles. In principle, condensation processes are reversible. In a hot universe, the heat that is released during condensation would flow back to a condensate, causing it to disintegrate shorty after its formation. Condensation products only remain stable when (1) the elements of the condensate do not react further to unstable products and (2) the reaction heat of condensation can be dispersed in a relatively cold environment (e.g., Lineweaver and Egan 2008). Stability of condensates can vary. For example, an isolated proton is extremely stable, while an isolated neutron has a half-life of approximately 15 min. The dispersal of heat, as a cause of stability of a condensate, is made possible because the universe expands. Expansion dilutes the energy (heat) in an ever-increasing volume, hereby lowering the temperature of the universe. Already soon after its beginning, the temperature of the universe was sufficiently low for quarks to form condensates of two or three quarks, called hadrons. With further lowering of temperatures, hadrons started to interact through pion exchange, causing low-complexity nuclides to form. Later, nuclear fusion in stars and supernova explosions allows for the formation of relatively large/heavy nuclides. And when the temperature sinks below 3000°K, nuclides and electrons condensate to atoms, which at still lower temperatures condensate to low-complexity molecules.

The formation of the above operators is not independent of the interaction systems surrounding them. Large interaction systems form when gravity pulls dispersed matter together, causing, e.g., stars, planets, and moons. Celestial bodies do not have dual closure, and are not operators for this reason. Celestial bodies offer the scaffolds for many condensation processes that lead to new operators.

As a wrap-up one can say that complexity from quarks to molecules, which are all *physical* operators, is made possible through condensation reactions. The heat of condensation is radiated into the colder environment of the expanding universe.

3.2 Group 2: Operators Based on Contained Autocatalysis, Causing Reproduction and Competition

In principle, one can view the cell as a product of a condensation reaction. After all, the formation of a lipid vesicle in an aqueous solution is a thermodynamically favorable event (Claessens et al. 2007; Fanelli and McKane 2008; Hernández-Zapata et al. 2009) as is the formation of complex molecules that can be found in early cells. But condensation is not the most relevant novelty of cells.

The most relevant innovation from a thermodynamic point of view is that a cell harbors a new kind of "machinery" named autocatalysis. Autocatalysis implies that a group of catalytic molecules in a solution cooperatively digest energy and chemicals from the environment and turn these into copies of the molecules that are already in the group. For a proper functioning of autocatalysis, it is necessary that the molecules are kept together, e.g., through a membrane. Without a membrane the molecules would disperse randomly until their concentration is too low for autocatalysis. Or there would be no individuality, because, without membranes, autocatalytic sets can mix. The membrane thus keeps the autocatalytic set together. In turn the autocatalytic set maintains the membrane through the production of membrane molecules.

The energetics of contained autocatalysis equals that of a water mill. A water mill is powered by the energy from the height gradient of the water before and after the mill. By analogy, the autocatalysis in a cell is powered by free energy gradients in the environment (Branscomb and Russell 2013). In principle, the rates of autocatalytic reactions are determined by the magnitude of external gradients. As a general rule, one can say that if more free energy from the environment is available for degradation, cells have more energy available for an increase in organizational complexity. In this way, properties such as maintenance and growth can be thermodynamically paid for through the degradation of external free energy gradients. Cells can now be viewed as degradation devices that have sprung from energy gradients. In the words of Lineweaver and Egan (2008): "food has produced us to eat it."

As long as sufficient environmental resources are available, a cell produces more new molecules than are necessary for balancing the losses. Molecules can be lost from the cell through degradation and/or reactions with other molecules. If there is net production, the numbers of the molecules in the cell increase: the cell grows. Cells that have grown larger may split automatically because of inherent thermodynamic stresses on the surface of the cell (Rashevky 1938; Corominas-Murtra 2019). Such splitting can be viewed as a primitive form of reproduction. In an environment that is sufficiently rich in resources, the splitting of cells will cause their numbers to increase exponentially. It is also relevant that the splitting of a cell offers a fundamental source of variation, because it is hard, again for thermodynamic reasons, to partition chemicals equally over the offspring cells.

The above explains how autocatalysis can lead to reproduction, and variable offspring, but does not yet explain the formation of organisms that are more complex than cells. After all, the production of a broad variety of cells does not automatically imply that larger or more complex cells will prevail. Actually, from a physical point of view, the opposite might be true. Small cells ingest more resources per volume, and grow faster, because they have a large surface-to-volume ratio. As they grow faster, and assuming that they divide at a small volume, small cells will outnumber larger cells. The laws of physics thus favor the production of small, primitive cells. This leads to the question of how more complex cells and organisms consisting of many cells could form.

To answer this question, one needs a cause, a "pressure," favoring larger and/or complex cells. As has been indicated by, e.g., Pross (2003), autocatalysis fuels growth and reproduction, which makes it a driving force that leads to crowding, which in turn leads to competition for space and resources. In a group of peers, cells that are better at gaining access or dominance over resources have better chances for survival and reproduction. The ability to gain access to resources, and to use them efficiently, has been discussed in the context of different strategies organisms can follow to increase their dominance over resources, or "resource dominance" (Jagers op Akkerhuis and Damgaard 1999). Resource dominance can be increased via several strategies. One major direction is complexity. To survive in a competitive environment and ascertain access to resources, while preventing being preyed upon, a cell needs extra features, which on average require more complex chemical pathways and/or special surface structures, flagella, a large size, etc. Competition between peers thus causes selection for complexity.

So far the discussion was about cells. As a next level in complexity, the OT points to combinations of cells, such as cells in cells, or adhering cells. The combination of a cell within a cell, classically named a eukaryote cell, is referred by the OT as "endosymbiont cell," that is, a cell with one or more endosymbionts. Examples of endosymbiont cells are the many different forms of protozoa, hosting, e.g., mitochondria or chloroplasts, or, as Bardele (1997) demonstrated, entire other protozoa. The causes leading to endosymbiont cells may differ, but as the leading theme one may think of a relative increase in the efficiency of resource use and resulting increase in resource dominance.

The combination of adhering cells is realized in multicellular organisms. The dual closure associated with multicellularity demands that cells are connected through plasma channels, allowing for functional closure, while sharing a common membrane, allowing for structural closure.

After nature produced cells, and endosymbiont cells, multicellular organisms did not evolve in one large step. Several mechanisms can have supported this transition. For example in blue-green algae, which are highly complex bacteria, the fixing of nitrogen in special cells initiated plasma channels between these cells and normal cells, a situation that classifies as bacterial multicellularity. An example of endosymbiontic multicellularity is offered by algae, where predatory interactions have selected for life cycles in which a single cell produces a clone of genetically identical daughter cells that remain attached (Herron and Nedelcu 2015; Herron et al. 2019). By analogy with a chemical polymer consisting of identical molecules, I suggest to refer to a clonal clump of cells as a *polycellular group*. As the cells are clonal, there is no reason for genetic conflicts between the individuals, and it is relatively unproblematic for ever-closer interactions to evolve, allowing, e.g., plasma channels and unicellular propagulation. According to the OT, the stage in a life cycle in which cells develop plasma channels classifies as a *multicellular organism*. Possibilities for the development of complexity are limited when a clump of cells harbor cells of different genetical origin, such as the slug of a slime mold. As the cells are not clonal, but have different genetical backgrounds, a lump of cells like these may be indicated as a *pluricellular group*. Pluricellular groups will generally exhibit genetic conflicts, which stand in the way of the evolution of close integration. Finally, if different species clump together, such as in Lychen, the group can be viewed as a symbiotic relationship. The terms polycellular group and pluricellular group are introduced to clearly describe different evolutionary pathways that may or may not lead to a multicellular organism.

The above suggests that, as a rule, competition favors complexity. But if this would be true, why have not all organisms on earth become complex? Several lines of reasoning can explain this. One explanation is that some complex organisms exploit small organisms as a resource, by preying on them. Strong predation will simultaneously reduce crowding of the prey, and favor prey with a high growth rate. Prey with high growth rates will no longer have much potential of developing complexity, simply because one cannot construct much complexity if one's generation time is very short and one's size is small. Besides predator-prey dynamics, there exist many other interactions explaining why not every organism is complex, e.g., living in small spaces, as many bacteria do, or living as a parasite, e.g., like wasps that lay their eggs in the eggs of beetles. It is interesting from a thermodynamic point of view that as the result of feeding relationships in food chains, more free energy will be degraded, as was corroborated by the work on the thermodynamics of model food chains by Meysman and Bruers (2013).

Competition between peers causes differential mortality of relatively unfit individuals, resulting in selection. It is thermodynamically relevant that selection "drives" evolution towards efficiency (e.g., Lotka 1922a, 1922b, 1945; Jagers op Akkerhuis and Damgaard 1999). This tendency is explained by Ulanowicz and Hannon (1987) as follows: "If two systems receive the same quantity of energy at the same entropy, that system which extracts the most work from its input before releasing it to its environment (as it inevitably must) can be said, in the second law sense of the word, to be the most efficient utilizer. Having extracted more work from the given amount of energy, the quality of the release is less, i.e., its entropy is higher." In other words, the more efficient degradation of free energy an organism has, the less time it takes to disperse a given amount of free energy, which tendency accords with the least action principle. More efficient use of resources will boost growth and reproduction. The population increases. And the overall degradation of free energy increases. This effect has a corollary in Jevons paradox in economy, where efficient production leads to cheaper goods, a strong increase in sales, and net increase in the exploitation of resource, instead of a decrease. At the same time, Sprengel (1839) and von Liebig (in 1840, in Salisbury 1992) already indicated that, like a chain that is as weak as its weakest link, the growth and performance of even the most efficient organisms are limited by the scarcest resource.

An aspect that was not yet discussed is the ratio between the net thermodynamic costs involved in the increase of the evolved complexity in organisms over generations (the increase in the genetics) versus the energy it takes to *evolve* more complex structures. In relation to this question Styer (2008) calculated that "at a minimum, the Earth is bathed in about one trillion times the amount of entropy flux required to support the rate of evolution assumed here." Styer (2008) continues: "the decrease in entropy required for evolution is so small compared to the entropy throughput that would occur even if the Earth were a dead planet, or if life on Earth were not evolving, that no measurement would ever detect it." Styer's calculations support that organisms degrade vastly more energy for safeguarding their maintenance, their growth, and their reproduction, than is needed for the actual mutations involved in complexity increase. However, such mutations, or other changes, would never have taken shape if there had not been the larger context of reproduction, competition, and mortality of the least fit individuals. In that context, it is furthermore logical to expect that organisms are not only selected for efficient use of resources, but that competition will also cause selection for increases in evolvability, as is discussed in the work of, e.g., Wagner and Altenberg (1996).

Summarizing the above, one can conclude that more complex, more efficient, and more evolvable organisms will prevail because of an interplay of several factors:

1. Autocatalysis taps energy from the environment. Sufficiently strong gradients force cells to grow. Growth necessitates division/reproduction, resulting in crowding and competition for resources.
2. Chance and thermodynamic "sloppiness" during autocatalysis offer the fundamental mechanisms behind the production of varying offspring.
3. Competition will with preference let "outperforming peers" survive and reproduce. Since outperformance is (on average) coupled to complexity, combining efficiency and extra features, competition favors complexity in part of the organisms.
4. Not all organisms become complex. An outperforming competitor, e.g., a predator, may suppress development towards complexity of its prey, or the environment may enforce size limits.
5. Due to competition, an organism must evolve towards efficient use of resources, and higher evolvability.
6. The free energy consumption involved in maintenance, growth, and reproduction, and the mortality of unfit individuals, vastly exceeds the free energy consumed by evolutionary increases in complexity.

A mechanism that allows complexity to increase in a major way is dual closure. Driven by competition, predation, cooperation, and other interactions that result in stress, or that release the organism from stress, and while proceeding one dual-closure step after the other, the complexity of part of the organisms on earth has increased from cells to endosymbiont cells (the "eukaryotes") and to multicellular organisms consisting of endosymbiont cells. The most recent dual closure in organisms takes the form of a (hypercyclic) neural network. Thermodynamic aspects of neural network organisms (called "memons" in the operator theory) are discussed in the next paragraphs.

3.3 Group 3: Operators Competing on the Basis of Neural Activity

As long as organisms lack neurons, they will predominantly compete for resources through their cellular or multicellular structure. For example in sessile organisms like trees, this leads to competition for photosynthetic area, and effective root surface, causing trees to become larger and broader, and roots to extend wider and/or deeper.

As soon as organisms have neurons, the coordination of motion allows additional means for resource competition. About 600 million years ago, nature enthusiastically experimented with the first primitive neural networks in combination with primitive, and often exotic, bodily designs. These experiments are witnessed by the many extraordinary fossils that suddenly appear in the (pre-)Cambrium. Later, behavioral competition intensifies, favoring power and efficiency. The Darwinian evolution of animals can be interpreted for a large part as a competition for neural network capacity. In some taxa, the capacity of the neural networks has increased steadily, to reach very high levels notably in animals with social behavior, such as dinosaurs, octopuses, crows, parrots, seals, dolphins, orcas, and humans.

3.4 Group 4: Artifacts Resulting from Design (and Competition)

There has been a marked increase in neural network capacity during the most recent 600 million years of Darwinian evolution. However, Darwinian evolution of neural network structure is relatively limited, because the production of every next generation takes time, because possibilities for genetic variation are limited, and because selection has low precision. While genes evolve per generation, the interactions between nerve cells in brains allow lifelong learning and thinking. Learning and thinking speed up complexity increase of every individual in a major way. Yet, learning too has limits, e.g., because education can take up to 15–40 years in humans, and because eventually, even the greatest minds die.

In the context of such impediments, it is relevant that a new evolutionary space has opened up, that of the artifacts, which are objects (both abiotic or biotic) that are at least minimally altered in their structure by humans or other intelligent species. Examples of artifacts are a stone axe, a hammer, a pig farm, a bicycle, a factory, an airplane, a city, or a computer. Artifacts can bypass the barriers of organismic evolution because they are purposefully constructed by man, and allow construction and modification by design, which can speed up complexity increase, as the following examples demonstrate.

The first primitive human artifacts, e.g., worked stones, stem from roughly 2.5 million years ago (Semaw 2000). Since that time, the complexity of artifacts has increased from stone axes, bows, and arrows to fishing nets, huts and houses, wheelbarrows, cars, airplanes, and computers. The operator theory predicts that in the near future we will most likely be able to design computers that, like humans, can not only learn a single task, such as playing chess or go, but also learn to master any task or activity, and for this reason are said to possess "general intelligence" (the operators involved are called "softwired memons," Jagers op Akkerhuis 2010b). One day generally intelligent computers may even have the ability of being aware of their existence, and thus be conscious (Willems 2012).

The trajectory from the first primitive organisms with neural networks to humans took 600 million years. The path from stone axes to intelligent computers, assuming that they will be constructed in a reasonably near future, will have taken about 2.5 million years. When talking about intelligent computers, one may think of a robot that successfully passes a stringent form of Turing test. A rough calculation now shows (600 million compared to 2.5 million years) that artifacts evolved more than 240 times as fast as did the neural networks of memons.

One can observe that, like complexity in organisms, the complexity of artifacts shows an increase in organization over time. Initially using stone plows, people soon developed iron plows. And iron plows became tractors, currently with GPS steering. Such developments demand a lot of trial and error, engineering, financial investments, and man-hours. It takes resources to push developments in the direction of GPS-steered tractors. As the driving force for this process, one must again think of the interplay of the necessity of humans to eat, and of societal competition. Competition in the socioeconomical context in which people develop artifacts does not predominantly refer to competition between individual people. Instead it must be viewed as taking place in a social/economic/scientific arena.

4 Discussion

A long-standing challenge in systems science is the question of how to define and quantify hierarchical complexity. For part of this challenge, and as was demonstrated in the current study, the operator theory can be used for relating hierarchical organization to causal processes and thermodynamics.

In principle the generality of the subject is an invitation to many discussions. Here I will focus on two discussion themes: (1) the surprising absence of a focus on kinds of systems in the hierarchy literature and (2) an analysis of the use of self-organization.

4.1 Can Consistent Hierarchies Be Constructed Without Paying Attention to Kinds of Systems?

In the introduction I have shown that the study of complexity hierarchies has a long history, ranging from, e.g., Feibleman (1954) to Laszlo (1972), Miller (1978), Jaros and Cloete (1987), Ebeling and Feistel (1992), Korn (2002), and Maynard Smith and Szathmáry (1995). Until recently, however, kind specificity of hierarchies had obtained little attention. But, if one does not pay attention to kinds, especially the "general" approaches to hierarchy risk ranking systems of dissimilar kinds. For example, a general construction rule such as "interactions between systems at level X produce a system at level X+1" cannot guarantee that any next system will be of the same logical kind. For example, if one starts with molecules, interactions between molecules can create a molecule, water, or a cell, as a next level. And starting with cells, interactions between cells can, as a next level, produce a lump of cells, an endosymbiontic cell (e.g., mitochondria in protozoa), or a population of cells. An unspecific/general construction rule thus allows for many different options, which causes problems due to kind inconsistency. For example, a hierarchy from molecules to cells and populations contains material systems (molecules, cells) as well as mental groupings (populations, see Fig. 2).

Overly general terminology has hindered closure research too. For example, in Turchin's metasystems theory (MST) the terminology (S, S', S_i) is not specific about the kinds of systems involved. As a consequence, MSTs, even when they focus on closure, do not distinguish different *kinds* of closures. And when Heylighen et al. (1995) in relation to MST theory introduces the term "supersystem," the examples offered are of different kinds: (1) The binding of quarks in a nucleon: This represents the functional aspect of the dual closure of the atom. (2) The binding of electrons: This represents the (complementary) structural aspect of the dual closure of the atom. (3) The binding of molecules in crystals: This depends on electrostatic bonding. Such bonding results in an interaction system (see Fig. 2). Only covalent bonding would comply with dual closure. (4) The bonding of planets through gravity: This results in an interaction system. Planets are aggregates, not operators. While the four examples count as transitions towards supersystems, the examples simultaneously demonstrate that the term "supersystem" is too general to serve as a basis for the construction of a hierarchy that is consistent in its kinds. For the specific purpose of creating a consistent complexity hierarchy, the focus of the OT on kinds of systems can be viewed as a fundamental innovation.

The above kind inconsistencies do not stand alone. In a broad analysis it was shown (Jagers op Akkerhuis 2016, Chap. 16.1.4) that kind inconsistency is present in many classical hierarchies, including those of Young (1976), Miller (1978), and Stikker (1992). And in Chaps. 8 and 10 of Jagers op Akkerhuis (2016), I analyze the major evolutionary transitions of Maynard Smith and Szathmáry (1995), Szathmáry and Maynard Smith (1995). Every major transition is based on the tryptich of reproduction as part of a larger unit, functional differentiation and changes in communication. Even though these criteria are rather restrictive, some transitions produce operators, others more complex elements inside operators, and still others compound objects, or mental groups (an overview of these terms is offered in Fig. 2). Consequently, the ranking of major transitions cannot result in a kind-consistent hierarchy.

Apparently, if one aims at creating a kind-consistent complexity hierarchy, the criteria need to be highly restrictive, and kinds must be part of the construction rule. The identification of highly restrictive criteria that allow kind consistency can be viewed as a fundamental contribution of the operator theory to system science.

4.2 *Self-Organization Versus Allo-Organization*

A central concept in complexity studies is self-organization. Self-organization refers to how a system organizes "itself." Here I pay attention to the nature of the word "self." Several different meanings of "self" were identified in the above paragraphs, and are discussed here in more detail.

The Self, and Self-Organization of Operators

When talking about operators, the idea of a "self" is supported by dual closure (Jagers op Akkerhuis 2019 Chapter 13). Dual closure offers a strong physical context for a self, because the functional closure and its interaction with the structural closure create higher order circularity, which implies clearly recognizable unity, and offers a context for self-reference. When talking about operators, dual closure furthermore creates a link between an operator as a physical unity and an operator as a conceptual level of organization.

Interactive Organization in Interaction Systems

The concept of "self" asks for a different interpretation in interaction systems. In contrast to an operator, an interaction system will generally represent a part of a larger environment from which it cannot be separated along a clear demarcation line. For example a river can be distinguished in a landscape, but it cannot be viewed independently of the landscape, nor can it "behave/operate" as a distinct individual. A river can be analyzed as the summed motion of many water molecules, and how these interact with the "riverbed," the groundwater, and the air. A well-known law for such "flow systems" is the constructal law of Adrian Bejan (1997). This law presumes an external factor imposing a current on the system, e.g., rain feeding a river, and states: "For a finite-size system to persist in time (to live), it must evolve in such a way that it provides easier access to the imposed (global) currents that flow through it." Accordingly, for the system to persist, it must stay the same or change in the direction of "easier access to the imposed currents." As is discussed by Bejan and Lorente (2010), the constructal law organizes and scrutinizes a range of thermodynamic optimality approaches.

The transport of material by a river has an analogy in the tree of life. Through reproduction, and selection, a pedigree of organisms forms. The flow of matter through the pedigree of organisms, generation after generation, and the interaction between organisms and the environment can be viewed as a generation-based river of individuals that interact with a "riverbank" represented by the ecosystem (e.g., Jagers op Akkerhuis 2016, pp 214, 229).

Allo-Organization of Artifacts

While it is general practice to speak about self-organization when talking about rivers, or cities, it is rather counterintuitive to use the term self-organization when talking about artifacts. The reason is that artifacts are constructed by humans. This implies that they do not organize themselves. For this reason, the improvement of the artifacts can be viewed as to be "scaffolded," a term used, e.g., by Godfrey-Smith (2009). Therefore, their production can be understood as externally scaffolded organization, or "allo-organization."

5 Conclusions

The literature offers several thermodynamic metrics that correlate with evolutionary complexity, such as energy rate density, volumetric Gibbs free energy, or action efficiency. The results of this study show that while these metrics apply to all systems, correlations with complexity can be made more specific if one uses the complexity ranking of the operators as a backbone.

The results of this study also bring to the fore that causal analyses, thermodynamics analyses, and organizational analyses highlight different aspects of the same reality. For example, a thermodynamic metric offers information about a relative state, or about tendencies, not about causes. And a causal analysis focuses on kinetics and processes, not on kinds of organization.

With the help of the operator hierarchy this study identifies four major groups of systems, each with their proper organization/complexity, causality, and thermodynamics. Amongst these groups are three major groups of operators (physical operators, organisms, memons) and the group of the artifacts.

For different reasons, neither an interaction system nor an artifact meets criteria that support the view that these systems organize themselves. Interaction systems are a product of the activities of operators in the system, while artifacts owe their construction to the willful design by memons. As a contribution to the way we view complex systems and understand change, such findings suggest a distinction between self-organization (of operators), interactive organization (of/in interaction systems), and allo-organization (of artifacts).

Acknowledgements The author likes to thank Dr. Toon van Eijk for commenting on a draft of this study.

References

Alvarez de Lorenzana JM (1993) The constructive universe and the evolutionary systems framework. In: Salthe SN (ed) Development and evolution. Complexity and change in biology. MIT Press, Cambridge, pp 291–308

Annila A (2010a) The 2nd law of thermodynamics delineates dispersal of energy. Int Rev Phys 4:29–34

Annila A (2010b) All in action. Entropy 12:2333–2358

Bak P (1996) How nature works: the science of self-organized criticality. Copernicus Press, New York, NY

Bak P, Sneppen K (1993) Punctuated equilibrium and criticality in a simple model of evolution. Phys Rev Lett 71(24):4083–4086

Bardele CF (1997) On the symbiotic origin of protists, their diversity, and their pivotal role in teaching systematic biology. Ital J Zool 64:107–113

Bejan A (1997) Advanced engineering thermodynamics. Wiley, New York

Bejan A (2016) The physics of life. The evolution of everything. St. Martin's Press, New York

Bejan A, Lorente S (2010) The constructal law of design and evolution in nature. Philos Trans R Soc B 365:1335–1347

Bernard C (1865) Introduction à l'étude de la médecine expérimentale, Paris

Blackmore S (1999) The meme machine. Oxford University Press, Oxford

Branscomb E, Russell MJ (2013) Turnstiles and bifurcators: the disequilibrium converting engines that put metabolism on the road. Biochim Biophys Acta Bioenerget 1827(2):62–78

Chaisson EJ (2001) Cosmic evolution: the rise of complexity in nature. Harvard University Press, Cambridge

Chaisson EJ (2011) Energy rate density. II. Probing further a new complexity metric. Complexity 17:44–63

Chandler JLR, de Vijver G (2000) Closure: emergent organizations and their dynamics. Ann N Y Acad Sci 901

Checkland P, Scholes J (1990) Soft systems methodology in action. John Wiley & Sons Ltd, Chichester

Claessens MMAE, Leermakers FAM, Hoekstra FA, Cohen Stuart MA (2007) Entropic stabilization and equilibrium size of lipid vessels. Langmuir 23(11):6315–6320. ISSN 0743-7463

Corominas-Murtra B (2019) Thermodynamics of duplication thresholds in synthetic protocell systems. Life 9:9. https://doi.org/10.3390/life9010009

Darwin C (1859) On the origin of species by means of natural selection, or he preservation of favoured races in the struggle for life. John Murray, London

Darwin C, Wallace AR (1858) On the tendency of species to form varieties; and on the perpetuation of varieties and species by natural means of selection. Zool J Linnean Soc 3:46–50

Dawkins R (1976) The selfish Gene. Oxford University Press, Oxford

Ebeling W, Feistel R (1992) Theory of self-organization and evolution: the role of entropy, value and information. J Non-equilibrium Thermodyn 17:303–332

Eigen M, Schuster P (1979) The hypercycle: a principle of self-organizalion. Springer, New York

Einstein A (1916) Die Grundlage der allgemeinen Relativitätstheorie. Annalen der Physik, vierte Folge Band 49:769–822

Fanelli D, McKane AJ (2008) Thermodynamics of vesicle growth and instability. Phys Rev E 78:051406

Feibleman JK (1954) Theory of integrative levels. Br J Philosophy Sci 5:59–66

Feigenbaum MJ (1978) Quantitative universality for a class of non-linear transformations. J Stat Phys 19:25–52

Georgiev GY, Chatterjee A (2016) The road to a measurable quantitative understanding of self-organization and evolution. In: op Akkerhuis GAJM J (ed) Evolution and transitions in complexity. The science of hierarchical organization of nature. Springer, Cham

Georgiev G, Georgiev I (2002) The least action and the metric of an organized system. Open Syst Inf Dyn 9:371–380

Georgiev GY, Henry K, Bates T, Gombos E, Casey A, Daly M, Vinod A, Lee H (2015) Mechanism of organization increase in complex systems. Complexity 21:18–28

Gibbs JW (1873) A method of geometrical representation of the thermodynamic properties of substances by means of surfaces. Trans Connecticut Acad Arts Sci 2:382–404

Gladyshev GP (1978) On the thermodynamics of biological evolution. J Theor Biol 75:425–441

Gladyshev GP (2017) Hierarchical thermodynamics: foundation of extended Darwinism. Imperial J Interdiscipl Res (IJIR) 3:1576–1588

Godfrey-Smith P (2009) Darwinian populations and natural selection. Oxford University Press, Oxford

Hernández-Zapata E, Martínez-Balbuena L, Santamaria Holek I (2009) Thermodynamics and dynamics of the formation of spherical lipid vesicles. J Biol Phys 35(3):297–308

Herron MD, Nedelcu AM (2015) Volvocine algae: from simple to complex multicellularity. In: Ruiz-Trillo I, Nedelcu AM (eds) Evolutionary transitions to multicellular life. Advances in marine genomics 2. Springer Science & Business Media, Dordrecht, pp 129–152. https://doi.org/10.1007/978-94-017-9642-2_7

Herron MD et al. (2019) Nature. Scientific Reports 9(2328)

Heylighen F (1990) Relational closure: a mathematical concept for distinction-making and complexity analysis. In: Trappl R (ed) Cybernetics and systems '90. World Science, Singapore, pp 335–342

Heylighen et al. (1995) (Meta)systems as constraints on variation. A classification and natural history of metasystem transitions. World Futures 45(1):59–85

Heylighen F (1999) The growth of structural and functional complexity during evolution. In: Heylighen F, Bollen J, Riegler A (eds) The evolution of complexity. Kluwer Academic, Dordrecht, pp 17–44

Heylighen, F., Joslyn, C. and Turchin, V. (1995) (eds.): The quantum of evolution. Towards a theory of metasystem transitions. Gordon and Breach Science Publishers, New York. (Special issue, vol. 45(1) of 'World Futures: the journal of general evolution'.)

Holling CS (1973) Resilience and stability of ecological systems. Annu Rev Ecol Syst 4:1–23

Hubble E (1929) A relation between distance and radial velocity among extra-galactic nebulae. PNAS 15(3):168–173

Jagers op Akkerhuis GAJM (2001) Extrapolating a hierarchy of building block systems towards future neural network organisms. Acta Biotheor 49:171–189. https://doi.org/10.1023/A:1011949326096

Jagers op Akkerhuis GAJM (2010a) The Operator Hierarchy, a chain of closures linking matter life and artificial intelligence, PhD thesis. Radboud University, Nijmegen

Jagers op Akkerhuis GAJM (2010b) Towards a hierarchical definition of life, the organism, and death. Found Sci 15:245–262

Jagers op Akkerhuis GAJM (2012) The pursuit of complexity. The utility of biodiversity from an evolutionary perspective. KNNV Publisher

Jagers op Akkerhuis GAJM (2014) General Laws and Centripetal science. Eur Rev 22:113–144

Jagers op Akkerhuis GAJM (2016) Evolution and transitions in complexity. The science of hierarchical organization of nature. Springer, Cham, Switzerland

Jagers op Akkerhuis GAJM (2018) Een realistische, trans-disciplinaire ontologie voor de empirische basis van de wetenschap: Bijdragen van de Operator Theorie. (A transdisciplinary realist ontology for the empirical basis in science: contributions of the Operator Theory). Researchgate https://doi.org/10.13140/RG.2.2.12851.12321

Jagers op Akkerhuis, G.A.J.M. (2019). ScienceBites – A fresh take on commonly used terms in science Wageningen Academic Publishers, 142 pp.
Jagers op Akkerhuis GAJM, Damgaard C (1999) Using 'resource dominance' to explain and predict evolutionary success. Oikos 87:609–614
Jagers op Akkerhuis GAJM (2008) Analysing hierarchy in the organization of biological and physical systems. Biol Rev 83:1–12
Jagers op Akkerhuis GAJM, van Straalen NM (1999) Operators, the Lego-bricks of nature, evolutionary transitions from fermions to neural networks. World Futures J Gen Evol 53(329-345)
Jaros GG, Cloete A (1987) Biomatrix, the web of life. World Futures 23:203–224
Kauffman SA (1986) Autocatalytic sets of proteins. J Theor Biol 119:1–24
Kauffman S (1993) The origins of order: self-organization and selection in evolution. Oxford University Press, Oxford, UK
Korn RW (2002) Biological hierarchies, their birth, death and evolution by natural selection. Biol Philos 17:199–221
Lambert FL (1999) Shuffled cards, messy desks, and disorderly dorm rooms – examples of entropy increase? Nonsense! J Chem Educ 76:1385–1387
Lambert FL (2002) Entropy is simple. Qualitatively. J Chem Educ 79:1241–1246
Lambert FL (2007) Configurational entropy revisited. J Chem Educ 84:1548–1550
Laszlo E (1972) Introduction to systems philosophy: towards a new paradigm of contemporary thought. Gordon and Breach, London
Lineweaver CH (2006) 6 cosmological and biological reproducibility: limits on the maximum entropy production principle. In: Kleidon A, Lorenz RD (eds) Non-equilibrium thermodynamics and the production of entropy. Understanding complex systems. Springer, Berlin, Heidelberg
Lineweaver CH, Egan CA (2008) Life, gravity and the second law of thermodynamics. Phys Life Rev 5:225–242
Lorenz EN (1963) Deterministic nonperiodic flow. J Atmos Sci 20(2):130–141
Lotka AJ (1922a) Contribution to the energetics of evolution. Proc Natl Acad Sci 8:147–151
Lotka AJ (1922b) Natural selection as a physical principle. Proc Natl Acad Sci 8:151–154
Lotka AJ (1945) The law of evolution as a maximal principle. Hum Biol 17:167–194
Mandelbrot BB (1967) How long is the coast of Britain? Statistical self-similarity and fractional dimension (PDF). Science 156(3775):636–638
Maupertuis de PL (1744) Accord de différentes loix de la nature qui avoient jusqu'ici paru incompatibles; Accord between different laws of nature that seemed incompatible (English translation)
Maynard Smith J, Szathmáry E (1995) The major transitions in evolution. W.H. Freeman Spektrum, Oxford
Meysman FJR, Bruers S (2013) Ecosystem functioning and maximum entropy production: A quantitative test of hypotheses. Phil Trans R Soc B 365:1405–1416
Miller JG (1978) Living systems. McGraw-Hill, New York
Moore, G.E. (1965). Cramming more components onto integrated circuits. Electronics. Retrieved 2011-08-22.
Moreno A, Mossio M (2015) Biological autonomy: a philosophical and theoretical enquiry. Springer, Dordrecht, the Netherlands
Newman MEJ (2005) Power laws, Pareto distributions and Zipf's law. Contemp Phys 46:323–351
Noether E (1918) Invariante Variationsprobleme. Nachr. D. König. Gesellsch. D. Wiss. Zu Göttingen, Math-phys. Klasse 1918:235–257.
Pagels HR (1985) Perfect symmetry: the search for the beginning of time. Simon and Schuster, New York
Penzias AA, Wilson RW (1965) A measurement of excess antenna temperature at 4080 mc/s. Astrophys J Lett 142:419–421
Pernu TK, Annila A (2012) Natural emergence. Complexity
Prigogine I, Stengers I (1984) Order out of chaos. Man's new dialogue with nature, New York
Prigogine I, Nicolis G, Babloyantz A (1972) Thermodynamics of evolution. Phys Today 25:23–28

Pross A (2003) The driving force for life's emergence: kinetic and thermodynamic considerations. J Theor Biol 220:393–406
Rashevky N (1938) Mathematical biophysics: Physico-mathematical foundations of biology. University of Chicago Press, Chicago, IL, USA
Salisbury F (1992) Plant physiology, 4th edn. Wadsworth, Belmont, CA
Scheffer M (2009) Critical transitions in nature and society. Princeton studies in complexity
Semaw S (2000) The World's oldest stone artefacts from Gona, Ethiopia: their implications for understanding stone technology and patterns of human evolution between 2·6–1·5 million years ago. J Archaeol Sci 27:1197–1214
Smith B (2004) Beyond concepts: ontology as reality representation. In: Varzi AC, Vieu L (eds) Formal ontology in information systems (FOIS), pp 1–12
Spier F (1996) The structure of big history: from the Big Bang until today. Amsterdam University Press, Amsterdam
Sprengel C (1839) Die Lehre vom Dünger oder Beschreibung Aller bei der Landwirthschaft gebräuchlicher vegetablilischer, animalischer und mineralischer Düngermaterialien, nebst Erklärung ihrer Wirkungsart, Leipzig
Stikker A (1992) The transformation factor. Towards and ecological consciousness. Element, Rockport, USA
Styer DF (2008) Entropy and evolution. Am J Phys 76:1031–1033
Szathmáry E, Maynard Smith J (1995) The major evolutionary transitions. Nature 374: 227–232
Teilhard de Chardin P (1959) The Phenomenon of Man. Harper, New York. (Original: Editions du Seuil I, Paris, 1955)
Teilhard de Chardin P (1969) The future of man. Collins, London. (Original: Editions du Seuil V, Paris, 1969)
Turchin VF (1977) The phenomenon of science, a cybernetic approach to human evolution. Colombia University Press, New York
Ulanowicz RE, Hannon BM (1987) Life and the production of entropy. Proc R Soc Lond B 232:181–192
Varela FJ (1979) Principles of biological autonomy. North-Holland, New York
Vogt L, Grobe P, Quast B, Bartolomaeus T (2011) Top-level categories of constitutively organized material entities. Suggestions for a formal top level ontology
von Bertalanffy L (1950) An outline of general system theory. Br J Phil Sci 1:134
von Bertalanffy L (1968) General system theory, foundations, development, applications. Penguin Books Ltd, Harmondsworth, Middlesex, England
Waddington CH (1969) Paradigm for an evolutionary process. In: Waddington CH (ed) Towards a theoretical biology, Vol 2: sketches. Edinburgh University Press, Edinburgh, UK
Wagner GP, Altenberg L (1996) Perspective: complex adaptations and the evolution of evolvability. Evolution 50:967–976
Willems, H.P. (2012). Why we need 'Conscious Artificial Intelligence'. https://mindconstruct.com/website/research
Young AM (1976) The reflexive universe. A Merloyd Lawrence book

Self-Organization in Stellar Evolution: Size-Complexity Rule

Travis Herman Butler and Georgi Yordanov Georgiev

Abstract Complexity Theory is highly interdisciplinary, therefore, any regularities must hold on all levels of organization, independent on the nature of the system. An open question in science is how complex systems self-organize to produce emergent structures and properties, a branch of non-equilibrium thermodynamics. It has long been known that there is a quantity-quality transition in natural systems. This is to say that the properties of a system depend on its size. More recently, this has been termed the size-complexity rule, which means that to increase their size, systems must increase their complexity, and that to increase their complexity they must grow in size. This rule goes under different names in different disciplines and systems of different nature, such as the area-speciation rule, economies of scale, scaling relations (allometric) in biology and for cities, and many others. We apply the size-complexity rule to stars to compare them with other complex systems in order to find universal patterns of self-organization independent of the substrate. Here, as a measure of complexity of a star, we are using the degree of grouping of nucleons into atoms, which reduces nucleon entropy, increases the variety of elements, and changes the structure of the star. As seen in our previous work, complexity, using action efficiency, is in power law proportionality of all other characteristics of a complex system, including its size. Here we find that, as for the other systems studied, the complexity of stars is in a power law proportionality with their size—the bigger a system is, the higher its level of complexity is—despite differing explosion energies and initial metallicities from simulations and data, which confirms the size-complexity rule and our model.

T. H. Butler
Biological and Physical Sciences Department, Assumption University, Worcester, MA, USA

G. Y. Georgiev (✉)
Physics Department, Assumption University, Worcester, MA, USA

Physics Department, Worcester Polytechnic Institute, Worcester, MA, USA
e-mail: ggeorgie@assumption.edu; ggeorgiev@wpi.edu; georgi@alumni.tufts.edu

G. Y. Georgiev, M. Shokrollahi-Far (eds.), *Efficiency in Complex Systems*,
Springer Proceedings in Complexity, https://doi.org/10.1007/978-3-030-69288-9_4

Keywords Self-organization · Stellar evolution · Size-complexity rule · Complex system · Principle of least action · Nucleosynthesis · Complexity theory · Quantity-quality transition

1 Introduction

This paper studies the phenomenon of self-organization in nature, its mechanisms, and connections to other characteristics of complex systems. The subject of quantity-quality transition dates back to Aristotle with the famous quote that "the whole is something besides the parts" (Aristotle 2018), but most of the Ancient Greek ideas they credited to have imported from Ancient Egypt, Babylon, Sumer, and possibly other ancient civilizations such as ancient India and China. Therefore, the origin of those ideas is possibly from prerecorded times, which proves that it has been tested again and again at different levels of understanding. It was given its name by Hegel who implies a causal relation between the quality and quantity of self-organizing systems, or a quantity-quality transition (Hegel 2014). More recently this has been termed a Size-Complexity rule (Bonner 2004). We apply our previous model in which each characteristic is proportional to each other characteristic of the complex system as it grows, not just the size and complexity, following power law dependencies (Georgiev et al. 2015, 2017). While there are so many principles and connections to discuss for different characteristics and different systems, which we will do in future papers, we will focus on only one—the size-complexity relationship—and its expression in nucleosynthesis during stellar evolution. Self-organization occurs only in systems far from thermodynamic equilibrium.

Complexity (quality), is an intensive property, which makes sense for a point or the parts of the system. Size (quantity), is an extensive property that describes the whole system. Size and complexity describe different aspects of self-organization, reinforcing each other in a positive feedback loop, as we see in our model (Georgiev et al. 2015). Self-organizing systems cannot exist without size and complexity causing an increase each other, with some exceptions for simpler systems. This is further shown in other work where Eric Chaisson has correlated the free energy rate density (FERD) in complex systems to their level of complexity and evolutionary stage, another size-complexity rule, and used the term Cosmic Evolution to describe this process (Chaisson 2002). The relevance of this connection to our work is that FERD is used to do the work to structure the system and build and maintain its complexity, an aspect which we will explore in future papers.

There are a variety of names for the size-complexity rule, as seen in the examples. The literature on size-structure relation is enormous, which confirms its proven and established rules and its universality. We only mention some examples. We just want to reiterate that the size-complexity rule is another term for quantity-quality transition, in evolving systems outside of thermodynamic equilibrium of physical, chemical, biological, or social nature. Their difference stems from the fact that they were rediscovered in different areas of science independently. They all study the same universal phenomenon in complex systems. This shows that this rule is identical in all systems and is universal, something that is a centerpiece of our research—to search for universality of complexity across systems of different nature.

This research has broader connection to other areas of science. In this paper, we study the progress of self-organization in stars as a function of their size. We use the degree of grouping of nucleons into elements by nucleosynthesis (progress of nucleosynthesis and therefore self-organization) as a measure of complexity. We chose this measure because it is clear and unambiguous, metallicity data are widely available for stars and galaxies by observations and simulations, can be related to other measures in other systems, and can be readily calculated from data. We calculate this progress in stellar evolution for stars of different masses and initial metallicities and with different explosion energies. The purpose of this is to test our model (Georgiev et al. 2015, 2017) and to compare it with the processes self-organization increase in other systems. We consider stars as complex systems, and one measure of their complexity is the degree to which they combine nucleons into heavier elements. The more of the heavier elements there are as a fraction of the mass of the star, the more advanced it is in its evolutionary stage and degree of complexity. There could be other measures, such as the differentiation of the internal structure of the star, but they will be related to this one. We chose simulations by Nomoto et al. (2006) of stars of different masses, explosion energies and metallicities at the end of their life, when they have exploded as supernovae. Those simulations were checked against observations of already exploded stars in the SAGA catalog where stellar composition can be measured by spectral analysis of their nebulae (Suda et al. 2008).

2 Stellar Evolution Overview

The energy from the Big Bang self-organized under the influence of the strong, weak, and electromagnetic forces into matter, with 75% of which was Hydrogen and the rest was primarily Helium, with trace amounts of Lithium. Eventually, large amounts of hydrogen atoms coalesced and created strong gravity centers, which forced these atoms together in a dense space and heated them up. This fused many of the hydrogen atoms together, releasing vast amounts of energy. The heat force opposes the force of gravity to keep a star in equilibrium and continues to add nucleons to existing elements to form new elements through nucleosynthesis, the process of self-organization we are studying. Stellar nucleosynthesis begins after the gravitational collapse of a dense, molecular cloud into a protostar. The mass of the protostar determines if it will reach the temperatures necessary for nuclear fusion and star formation. Towards the end of a star's life, it inefficiently forms heavier elements beyond iron, which absorbs heat and energy instead of releasing it. Eventually, the star no longer has sufficient heat to oppose gravity, and it collapses in on itself and explodes (Thielemann et al. 2018). During this explosion, heavier elements, such as gold and platinum, are synthesized. Therefore, the complex systems that are being examined in this paper are stars synthesizing elements throughout the course of their lifetime, including when they explode in a supernova event. The elemental abundances can then be detected using spectrophotometry.

2.1 *Other Stellar Studies and Models*

Other research has been done on stars to see how their nucleosynthesis and supernova event describes the chemical evolution of our galaxy using observational data of supernovae and metal-poor stars (Nomoto et al. 2013). Similar research has been done on how these yields (Nomoto et al. 2006, 1997, 2013) are affected by hydrodynamic effects during hypernova and supernova explosions (Nomoto and Suzuki 2013). Moreover, studies have been done on nucleosynthesis to accurately understand the abundance pattern of Pop III stars leading to hypernovae (Nomoto 2016). In addition, nucleosynthesis yields have been used to distinguish high-density Chandrasekhar-mass models and lower-density white dwarfs (Mori et al. 2018). These studies are important because they show patterns of chemical evolution and nucleosynthesis (Nomoto et al. 2013; Nomoto 2016). The degree of grouping of nucleons as a definition for complexity is inversely proportional to nucleon entropy which decreases during nucleosynthesis (Avellar et al. 2016).

Later simulations utilized the heavy-flavor neutrinos emitted from proton stars, like SN1987A, for triggered parameter explosions to more accurately report chemical evolution and iron group nucleosynthesis yields of in proto-neutron stars, taking into consideration the electron fraction of the ejecta (Sinha et al. 2017; Curtis et al. 2018). Another simulation has been done to look at the stellar yields of the first supernovae in stars of 12–140 solar masses and how rotation affects the nucleosynthesis yields (Takahashi et al. 2014). Simulations on nucleosynthetic yield for asymptotic giant branch, white dwarf, and core collapsing stars have been performed (Ritter et al. 2018; Pignatari et al. 2016). We find that the model of Nomoto 2006 is the best for studying nucleosynthesis because it shows the abundances of isotopes of various metallicities and solar massed stars as suggested by other reports (Wanajo et al. 2009). More recent studies of nucleosynthesis have not shown the chemical abundances of isotopes when initial metallicities and explosion energies vary. These results are useful in studying the size-complexity rule amongst stars of varying size, initial metallicities, and explosion energies. Many of the recent papers above have discussed abundance levels in different simulated stars. However, none have reported their star's yields as thoroughly as Nomoto.

2.2 *Hypothesis*

Earlier research has shown that the size-complexity rule is valid in a variety of non-equilibrium thermodynamic systems, as evident in biological cells that form spherical structures (Amado et al. 2018; Bell and Mooers 1997). Research has been conducted to investigate the size-complexity in the life cycle (Bonner 2015, 1995). In general, these studies demonstrate that there are two ways to measure complexity increase. One is the differentiation as a function of size, and the other is the structure formation (integration) in the larger system as a result of this differentiation. For example, the number of different cells in an organism is one measure, but it is correlated to the structure formation in organs and the overall functioning of the organism. There are rare exceptions, such as of slime mold which can consist of one cell, but the vast majority of organisms follow the general rule (Bonner 2004; McCarthy and Enquist 2005).

In stellar evolution, nucleosynthesis produces atoms that are different than those that initially existed, and in general, the larger a star is, the more variety of atoms it can produce. This, in turn, works in a manner analogous to complexity in biological systems, with larger systems having more internal differentiation, which for stars with a large variety of elements determines layered regions inside the star of different density, temperature, kinds of atoms, and nuclear reactions occurring there. The differentiation leads to change in the global overall structure in the system, as observed in stars, organisms, cities, economies, etc. We argue that the existing physics laws, with extension and modification, are sufficient to describe self-organization in all those different systems (Walker 2019).

Thus, our hypothesis, based on our previous model and data, Georgiev et al. (2015), Georgiev et al. (2017) is that larger stars will have greater progress of nucleosynthesis, complexity of element structure, at the end of their lives. This is because higher mass stars are much hotter, and the gravitational force is greater, causing greater pressure and density of matter at the core, which allows them to fuse more nucleons in their shorter life.

This has analogs in other complex systems, based on the size-complexity rule, which states that a system's complexity is contingent on its size. As we noted in our previous publications on Core Processing Units (CPUs) evolution, the level of organizational complexity and size were found to be locked in a positive feedback loop, and consequently, both increase exponentially through time, as a power law function of each other (Georgiev et al. 2015, 2017). Our research aims to determine whether there is a similar trend in stars undergoing nucleosynthesis or not.

3 Theory

3.1 Model and Overview of Previous Work

Here we include for reference the basic model which studies how two of the characteristics (observables, descriptors, measures, properties): size and complexity depend on each other and what time behavior can be predicted in general for complex systems (Georgiev et al. 2015). In this previously published paper, Georgiev et al. (2015), we labeled α as a measure of complexity, or level of organization, and Q is the measure of the size of the system. The measure of complexity α is the average action efficiency per one event in the system based on the Principle of Least Action (for any event in the universe, action tends towards a least value) which in this model drives self-organization forward and Q is the total amount of action of the system (Georgiev and Georgiev 2002; Georgiev et al. 2015, 2017; Georgiev and Chatterjee 2016; Georgiev et al. 2016). In this paper, the analogs to those measures are progress and mass. To achieve higher complexity, the system needs to have larger size. This model explores a positive feedback loop (reinforcing) between the quality and quantity, or size and complexity (Georgiev et al. 2015). It is supported by the observations about the size-complexity rule (Bonner 2004; Carneiro 1967; Bell and Mooers 1997) and size-efficiency rules (Bejan et al. 2011; Kleiber et al. 1932; West et al. 1999). An increase in an extensive property of the size of a system, drives it further away from equilibrium. This allows more work to be done to organize, leading to higher levels of organization reflected in its action efficiency per one event, α as the numerical measure for organization. Here we only present the results that are relevant to this application of the above-developed model. The solutions of the model are an exponential growth of quantity and quality in time, and a proportionality between them, which obeys a power law equation. In this paper, we study only the power law proportionality of quality from quantity, as the data presented do not have time resolution with the lifetime of the star, and the size of stars are fixed by the primordial gas cloud. Therefore, the size cannot grow in response to the increase of the complexity of the star, which is one exception of our model, because stars are simpler systems than biological and social. Nevertheless, stellar systems obey the size-complexity rule, as their structure is dependent on their size, as it will be seen in the results of this paper.

The exponential growth in time for the model is verified by data for CPUs (Georgiev et al. 2015). Both quantities, α and Q, fit very well with exponentials on log-linear plots. Eliminating time, a solution of the exponential equations is the power law relation (Eq. 1).

$$\alpha = \eta \cdot Q^{\gamma}, \tag{1}$$

where γ and η are constants coming from the solutions of the exponential equations (Georgiev et al. 2015).

Figure 1 displays the numerical solution of the power law proportionality prediction of our model, shown in Eq. 1 (Georgiev et al. 2015), for how the quality of stars in terms of progress of nucleosynthesis and internal organization and nucleon entropy reduction will depend on size, on a log-log plot. Here, for brevity, we just show one dependency from all possible extensive measures of size, (such as total free energy rate density, total power, total number of nucleons, etc.), namely mass. Therefore, in this paper, we test whether stars obey the prediction that their complexity at the end of their life is dependent on their mass, as a power law. Our full model predicts that the progress of nucleosynthesis will be exponential during the stellar lifetime, but there are no data yet to test this prediction. However, since it

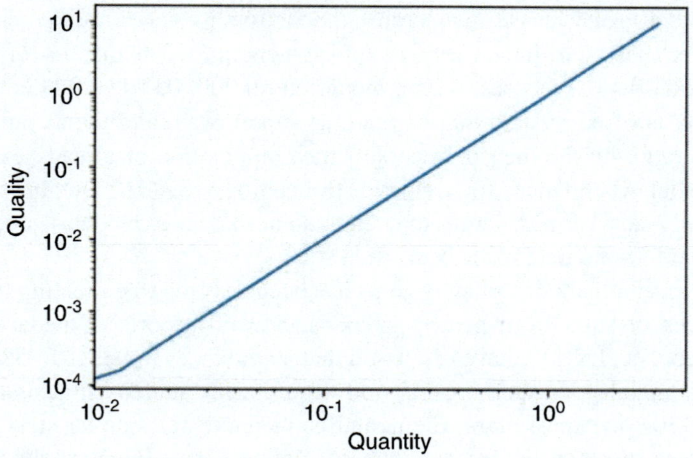

Fig. 1 Numerical solution of quality (organization, complexity) vs. quantity (size of the system) from the system of differential equations of our model, in arbitrary units

is connected to the power law solutions, confirming them will strengthen the model and, therefore, the exponential growth prediction.

The goal of this continuous study is to find universality—mechanisms that are unchanged across the broadest range of systems of different nature, having explanatory power for all of them. The size-complexity rule is only one of the dependencies in the model of self-organization (Georgiev et al. 2017). Each characteristic of a complex system depends on all others, not just the size and complexity, by a power law. In the general model, quantity accumulation leads to quality increase, but also, quality increase leads to quantity accumulation, which is the case in more advanced systems, such as biological and social, but not in Stellar Evolution.

4 Data and Methods

4.1 Data and Previous Simulations

The data represent the nucleosynthesis yields as a function of initial metallicity and stellar mass from nucleosynthesis yields of core collapse supernovae and galactic chemical evolution. We use it to check whether the progress of nucleosynthesis obeys a power law as a function of size, as shown in the previous model (Georgiev et al. 2015). These yields are based on the new developments in the observational and theoretical studies of supernovae and extremely metal-poor stars in the halo, which have provided excellent opportunities to test the explosion models and their nucleosynthesis (Nomoto et al. 2006). In this paper, the initial metallicities of the stars studied are 0.000, 0.001, 0.004, and 0.02. Their masses are 13, 15, 18, 20,

25, 30, and 40 solar masses, when only the metallicity varies. Additionally, we also study when these stars have varying explosion energies. The masses for these are 20, 25, 30, and 40 solar masses, with a metallicity 0.000, 0.001, 0.004, and 0.02. This simulation is of large stars that explode as supernovae, and their composition can be compared with the measurements of the composition of already exploded stars in the Stellar Abundances for Galactic Archaeology (SAGA) Database (Suda et al. 2008). Only stars of mass larger than eight solar masses explode as supernovae, that is why there are no data for less massive stars.

Much research about stellar systems has been devoted to collecting the elemental abundances of stars from nearby galaxies. Several reports (Umeda and Nomoto 2003; Frebel et al. 2005) have discussed that abundances in HE0107-5240 and other extremely metal-poor stars are in good accord with nucleosynthesis that happens in 20 to 130 solar mass stars. Elemental abundances for smaller stars, such as red giants, were reported in Omega Centauri (Johnson and Pilachowski 2010). Other research (Wanajo et al. 2009) has studied the yield and nucleosynthesis of unstable elements and reported their abundance for ST and FP3 model stars.

Further simulations (Tominaga et al. 2014) presented Pop III SN models, in which nucleosynthesis yields individually reproduce the abundance patterns of 48 metal-poor stars. Observations of abundances found in extremely metal-poor stars, HE 1300+0157, have also been done (Frebel et al. 2007). Another study (Prantzos et al. 2018) shows evolution of abundance of elements from Hydrogen to Uranium occurring in the Milky Way halo through a chemical evolution model of metallicity dependent isotopic yields from large stars. Some chemical abundances of extremely metal-poor stars from Pop III stars have been shown to describe the nature of first-generation stars formed after the Big Bang (Nomoto et al. 2005; Steigman 2007).

Abundance information is critical to our research because, using stellar calculations (Nomoto et al. 2006) based on known abundances (Suda et al. 2008), we determined the progress of nucleosynthesis of elements at the end of the stars' life. Other investigators' search for abundances of exploded and simulated stars allows us to apply our model to their findings and see how efficient nucleosynthesis is in both massive and small-scale stars.

4.2 Methods

The stellar yields of various isotopes, ranging from ^{1}H to ^{71}Ga, were taken from 13, 15, 18, 20, 25, 30, and 40 solar mass stars with varying metallicities: 0, 0.001, 0.004, and 0.02. These yields of each isotope were given in solar masses from the SAGA Database Suda et al. (2008) and Nomoto et al. (2006). From the raw data of Nomoto et al. (2006), the number of solar masses of each isotope from ^{1}H to ^{71}Ga and elements heavier than ^{71}Ga present within each star at the end of its life was first converted to the total number of nucleons present within each star as a measure of its size. In our calculations of progress of nucleosynthesis, we excluded the Hydrogen and Helium isotopes that existed before the star was formed when calculating the progress of nucleosynthesis. To determine the amount of Helium produced by the stars, we used Eq. 2:

$$He_{nuc}[M_{\odot}] = \left(\frac{\psi}{M_*} - \frac{25.2}{100}\right) \cdot M_*. \quad (2)$$

(All of the symbols are listed in the footnote.)[1]

The total amount of Helium in each star was summed and divided by the total number of solar masses to obtain its fraction from the stars' mass. Then the fraction of Helium originally present immediately after the Big Bang was subtracted from it to find the fraction of the star made of synthesized Helium by the star in solar masses (Wagoner et al. 1967). This number is reported as He_{nuc}. For this fraction, we used a mass number of four, because that isotope is the majority of Helium present in the star.

To more accurately calculate the progress of nucleosynthesis, we also included the elements heavier than ^{71}Ga even though Nomoto (Nomoto et al. 2006) excluded them, because those would have only been measured in trace amounts. When the estimates of all of those isotopes are added together, they form a significant fraction of the masses of the stars. Because information for these elements was not provided, the mass number of these is assumed to be 140, because that is approximately the mean mass number between Gallium and Uranium in the periodic system. These elements are referred to as "$^{140}\chi$". The total mass of $^{140}\chi$ was calculated by subtracting from the total initial mass of the star the ^{1}H to ^{71}Ga, the equivalent mass fraction of the explosion energy, and M_{cut}, all in the same units of solar masses. To find the mass of the explosion energy, the following equation was used (Einstein 1905):

$$E = M \cdot c^2, \quad (3)$$

where E is energy in joules, M is the mass in kilograms, and c is the speed of light in vacuum. This mass was then converted to solar masses by dividing it by the mass of the sun in kilograms.

[1] Symbols: $He_{nuc}[M_{\odot}]$ is the mass of the helium nucleosynthesized in the star in solar masses. n distinguishes the isotopes of all elements. β_n is the total number of nucleons for each isotope, n, within the star. β_i is the total number of nucleons present in each star. $M_{\odot}$ is a unit for the number of solar masses. $M_{\odot}[kg]$ is the mass of the Sun in kilograms. M_{is} is the mass of each isotope present in the star in solar masses. M_{cut} is the total mass of the remnant after the explosion of the supernova at the center of the nebula in solar masses. N_A is Avogadro's number. ψ is the combined solar masses of Helium-3 and Helium-4. M_* is the mass of a star in solar masses. $^{140}\chi$ represents the mass of the elements more advanced than ^{71}Ga in solar masses. $^{100}\rho$ is the mass of the heavier than helium elements that were made by a previous generation star (based on its metallicity). ϵ_n is the degree of grouping of nucleons for a selected individual isotope. A is the mass number of a selected isotope. ϵ_{sum} is grouping of all nucleons in each star. P is the progress of nucleosynthesis, the grouping of all nucleons in each star, normalized by its size (total number of nucleons).

Additionally, we do not integrate the M_{cut} mass value of Nomoto et al. (2006) into our calculation for progress of nucleosynthesis because of the lack of information about its exact composition. This is due to the fact that the material in the nebula contains information about all elements present in the star, but not the ones remaining in the central remnant object. The figures and trend line calculations were plotted and fitted using Python Spyder version 3.6.1.

4.3 Initial Metallicity

In the following calculations, we took into account the initial metallicity of the star to exclude those heavier than Helium elements that were synthesized by a previous star. Because these specific elements are not listed, we assume that they have a mass number of 100 and factor them into these calculations as "$^{100}\rho$." 100 is the approximate mean mass number of all of the elements in the periodic system, excluding hydrogen and the pre-synthesized helium, taking into account their relative abundance. To calculate the number of solar masses of $^{100}\rho$, the initial metallicity of each star as a fraction of the stellar mass was multiplied by its M_*, the star's mass in solar masses.

To find the progress of nucleosynthesis, we first calculated the number of nucleons of each isotope present in the star at the end of its life using Eq. 4 for each studied star.

$$\beta_n = M_\odot[kg] \cdot M_{is}[M_\odot]_n \cdot 10^3 \left[\frac{g}{kg}\right] \cdot N_A. \tag{4}$$

The number of nucleons, β, for each isotope, n, for the stars with metallicities: 0, 0.001, 0.004, and 0.02, is shown in Tables 8, 9, 10, and 11, respectively, in Appendix.

We calculated from its mass the total number of nucleons present in each star, β_i, which is a measure of its size, with Eq. 5:

$$\beta_i = M_\odot[kg] \cdot M_*[M_\odot] \cdot 10^3 \left[\frac{g}{kg}\right] \cdot N_A. \tag{5}$$

The total number of nucleons, β_i, for stars of different masses are shown in Table 1. We then multiplied the number of nucleons of each isotope for synthesized

Table 1 Initial total number of nucleons of each star

$M_*(M_\odot)$	13	15	18	20	25	30	40
β_i	1.56×10^{58}	1.80×10^{58}	2.16×10^{58}	2.40×10^{58}	2.99×10^{58}	3.59×10^{58}	4.79×10^{58}

Helium, and for all isotopes from ^{6}Li to ^{71}Ga, and $^{140}\chi$ by its mass number, which is the number of nucleons of each individual isotope, to obtain ϵ_n as a measure of the degree of grouping of the nucleons, using Eq. 6:

$$\epsilon_n = \beta_n \cdot A, \tag{6}$$

where A is the mass number of the isotope. To get ϵ_{sum} (Eq. 7), we took the ϵ_n of each isotope of synthesized Helium, ϵ_{He}, the sum of all isotopes from ^{6}Li to ^{71}Ga, $\sum_{n=6}^{71} \epsilon_n$, the number of nucleons of $^{140}\chi$, ϵ_χ, and added them together. From this number we subtracted the number of nucleons of $^{100}\rho$, ϵ_ρ to avoid including isotopes that were not synthesized by the star. The mass of M_{cut}, explosion energy, and the preexisting Hydrogen and Helium in the star do not participate in this calculation.

$$\epsilon_{sum} = \epsilon_{He} + \sum_{n=6}^{71} \epsilon_n + \epsilon_\chi - \epsilon_\rho. \tag{7}$$

The level of progress (complexity) for how far stars went in grouping nucleons together into heavier isotopes over their lifetime was determined. The more connected the nucleons are, the more advanced the nucleosynthesis is in terms of the degree of complexity and progress of filling the periodic system by that star. The progress of nucleosynthesis for each star, P, is then determined by dividing ϵ_{sum} by the total number of nucleons in stars 13, 15, 18, 20, 25, 30, and 40 $M_\odot$ (Eq. 8):

$$P = \frac{\epsilon_{sum}}{\beta_i}. \tag{8}$$

The progress of nucleosynthesis was determined for each star when the metallicity fraction equals 0, 0.001, 0.004, and 0.02 and reported in Table 2.

Figure 2 shows the progress of nucleosynthesis versus the initial number of M_* with initial metallicities equal to 0, 0.001, 0.004, and 0.02.

Table 2 Progress of nucleosynthesis within each star when the metallicity varies, subscripts next to P indicate the metallicity of stars in the row

$M_*(M_\odot)$	13	15	18	20	25	30	40
P_0	1.63	2.20	2.85	3.56	3.64	4.26	5.34
$P_{.001}$	2.44	2.12	3.06	5.41	7.42	9.16	12.49
$P_{.004}$	2.72	6.88	12.64	5.17	7.98	1.550	29.40
$P_{.02}$	2.54	7.40	9.78	11.93	20.24	26.75	66.15

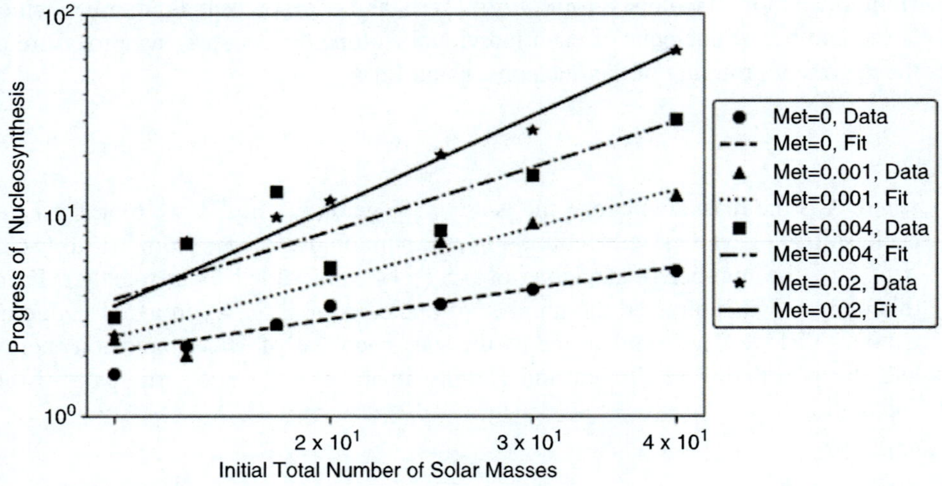

Fig. 2 Progress of nucleosynthesis vs. mass on a Log/Log scale, indicating a power law proportionality, confirming our model

4.4 Varying Explosion Energies

The stellar yields of various isotopes, ranging from ^{1}H to ^{71}Ga, were taken from 20, 25, 30, and 40 solar massed stars with varying explosion energies and metallicities of 0, 0.001, 0.004, and 0.02. Stars with 20 and 25 solar masses have 10 E of explosion energy, where in all cases, $E \sim 1x10^{51} ergs$. 30 solar massed stars have 20E of explosion energy. Stars of 40 solar masses have explosion energies of 30E. These yields of each isotope were given in solar masses from the SAGA Database Suda et al. (2008) and Nomoto et al. (2006). The method used for determining the P of varying metallicities stars is the same for when the metallicity and explosion energy vary. Table 3 shows the ϵ_{sum} of each star when the explosion energy varies and the metallicity equals 0, 0.001, 0.004 and 0.02. The progress of nucleosynthesis is reported in Table 4.

5 Results

In this section, we present a study of stars with varying initial metallicities, without taking into account the explosion energies. Tables 8, 9, 10, and 11 in Appendix show the calculated number of nucleons for each isotope present in stars of various masses (Nomoto et al. 2006), when the initial metallicities equal 0, 0.001, 0.004, and 0.02, respectively. Table 1 shows the total number of nucleons of each star, assumed constant throughout its life.

Table 3 ϵ_{sum} of each star when explosion energy varies, subscripts next to ϵ_{sum} indicate the metallicity of stars in the row

M_* ($M_\odot$)	20	25	30	40
E (10^{51} $ergs$)	10	10	20	30
ϵ_{sum0}	7.75×10^{58}	8.28×10^{58}	1.26×10^{59}	2.01×10^{59}
$\epsilon_{sum.001}$	5.44×10^{58}	1.03×10^{59}	2.22×10^{59}	3.87×10^{59}
$\epsilon_{sum.004}$	1.35×10^{59}	2.05×10^{59}	7.31×10^{59}	1.57×10^{60}
$\epsilon_{sum.02}$	2.78×10^{59}	6.05×10^{59}	9.51×10^{59}	3.20×10^{60}

Table 4 Progress of nucleosynthesis in stars when metallicity and explosion energy varies, subscripts next to P indicate the metallicity of stars in the row

M_*($M_\odot$)	20	25	30	40
E (10^{51} $ergs$)	10	10	20	30
P_0	3.24	2.77	3.50	4.20
$P_{.001}$	2.27	3.43	6.18	8.07
$P_{.004}$	5.63	6.83	20.35	32.80
$P_{.02}$	11.603	20.188	26.478	66.693

Table 5 ϵ_{sum} of each star when only Metallicity varies, subscripts next to ϵ_{sum} indicate the metallicity of stars in the row

M_* ($M_\odot$)	13	15	18	20	25	30	40
ϵ_{sum0}	2.53×10^{58}	3.95×10^{58}	6.14×10^{58}	8.53×10^{58}	1.09×10^{59}	1.53×10^{59}	2.56×10^{59}
$\epsilon_{sum.001}$	3.80×10^{58}	3.82×10^{58}	6.59×10^{58}	1.30×10^{59}	2.22×10^{59}	3.29×10^{59}	6.01×10^{59}
$\epsilon_{sum.004}$	4.23×10^{58}	1.24×10^{59}	2.72×10^{59}	1.24×10^{59}	2.39×10^{59}	5.57×10^{59}	1.41×10^{60}
$\epsilon_{sum.02}$	3.95×10^{58}	1.33×10^{59}	2.11×10^{59}	2.86×10^{59}	6.06×10^{59}	9.61×10^{59}	3.17×10^{60}

Table 5 shows the ϵ_{sum} of stars of each mass when their metallicity varies. Table 2 shows the progress of nucleosynthesis when the initial metallicity is 0, 0.001, 0.004, and 0.02, respectively. Figure 2 shows the progress of nucleosynthesis versus the number of solar masses for stars of each mass and metallicity on a Log/Log scale at the end of the stars' life. This shows that the progress of nucleosynthesis follows a power law in stars even when their initial metallicities vary. The advance in the progress of nucleosynthesis for each metallicity is similar as the trend lines show. The significance that they follow a power law is that it matches our previous model of interdependence of the characteristics of complex systems and the power law proportionality between them, which is confirmed empirically by all data for size-complexity rules by Bonner, Carneiro, and others (Bonner 2004; Carneiro 1967; Georgiev et al. 2015, 2017). We use this study on the one hand to additionally test empirically our previously published model, and on the other hand to apply that model to learn more about stellar evolution and to be able to extrapolate it to predict how larger or smaller stars will evolve based on what we have learned from stars between 13 and 40 solar masses. We are trying to gain the largest generality of the model and also use what we learn from stars to apply it to other complex systems as listed elsewhere in the text. The predictions of our previous model are for power law

Table 6 Coefficients and powers for the progress vs. solar mass trend line of Fig. 2

Metallicity	Coefficient	Power
0	2.28×10^{-1}	0.86
0.001	5.74×10^{-2}	1.47
0.004	3.71×10^{-2}	1.80
0.02	4.48×10^{-3}	2.60

Table 7 Coefficients and powers for the progress vs. solar mass trend line of Fig. 3 when explosion energies vary

Metallicity	Coefficient	Power
0	6.42×10^{-1}	0.50
0.001	1.88×10^{-2}	1.65
0.004	3.37×10^{-3}	2.50
0.02	2.61×10^{-3}	2.75

relationships, and the data for stellar evolution from simulations done by Nomoto support those predictions, which further strengthens the model. This increases our level of confidence in the model and in its applicability for other systems.

Table 6 shows the coefficients, represented by c, and the powers, represented by y, for the fits on Fig. 2, where the progress of nucleosynthesis is a function of the initial number of nucleons. These equations follow the format:

$$P = c \cdot M_*[M_\odot]^y. \tag{9}$$

Within the data of this simulation, there is an additional trend that the higher the initial metallicity, the greater the increase in progress as a function of mass. As it can be seen in Table 6, the powers of the fits increase from 0.86 to 2.6 for stars from 0 initial metallicity to stars of 0.02 initial metallicity.

5.1 *Varying Explosion Energy, Varying Initial Metallicity*

Table 4 shows the progress of nucleosynthesis within each star when explosion energies vary and the initial metallicity is 0, 0.001, 0.004, and 0.02, respectively. Figure 3 shows the progress of nucleosynthesis versus the mass of stars in solar masses for each of the given metallicities at their explosion energies on a Log/Log scale at the end of each star's life. This shows that the progress of nucleosynthesis follows a power law in stars even when their initial metallicities and explosion energies vary, as the trend lines show.

Table 7 shows the coefficients and the powers of the equations of Fig. 3 where the progress of nucleosynthesis is function of the stellar mass. These equations follow the format of Eq. 9.

We observe that the progress of nucleosynthesis is increasing with the size of the star as measured by its mass, which confirms the size-complexity rule, as reported by other authors (Bonner 2004; Carneiro 1967). Within the data of this simulation,

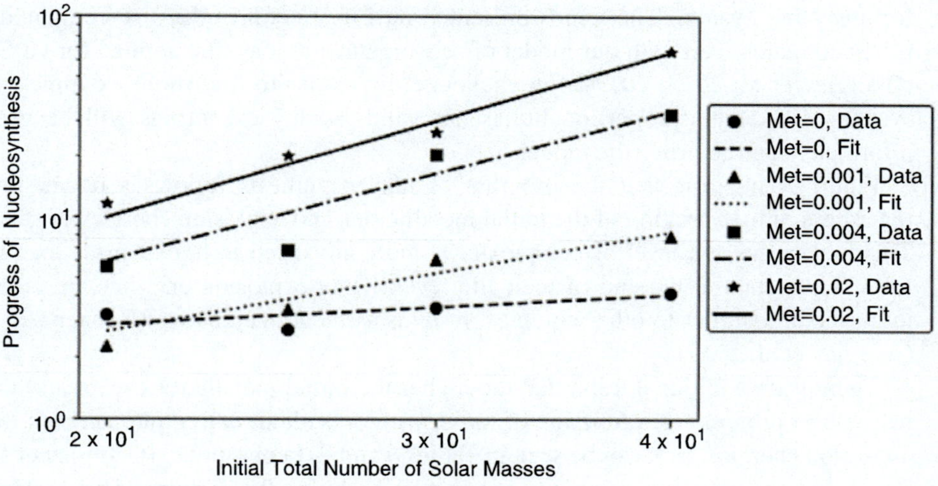

Fig. 3 Progress of nucleosynthesis vs. mass on a Log/Log scale when the explosion energies vary

there is an additional trend that the higher the initial metallicity, the greater the increase in progress as a function of mass. As it can be seen in Table 7, the powers of the fits increase from 0.5 to 2.75 for stars from 0 initial metallicity to stars of 0.02 initial metallicity.

There are a total of only four points in the graphs describing trends in stars with varying explosion energy, compared to the seven points in graphs where explosion energy is not taken into account. This is because (Nomoto et al. 2006) exclude 13–18 solar massed stars from their yields tables when the explosion energy varies.

6 Discussion

Figure 2 shows that despite the variation in the initial metallicity of the star, its progress of nucleosynthesis at the end of its life increases when the size of the star increases. This means that the system's size is a determinant of its level of self-organization. This can be compared to the size-complexity rule in all other self-organized systems of any nature—the number of ant castes as a function of the size of an ant colony, the number of different types of cells as a function of the size of an organism (Bonner 2004), the number of organizational traits in societies as a function of the size of those societies (Carneiro 1967), the species diversity as a function of the area size of the ecosystem, the number of species as a function of the size of a population and the size of the area that they occupy (Maurer 1996), the number of occupations in a human society as a function of the number of people (Bonner 2004), the GDP per person in cities as a function of city size (West 2017), the action efficiency of CPUs (Georgiev et al. 2015) and almost any other example for any other system. The trends presented on Fig. 2 confirm the size-complexity rule, and match well with our model of self-organization, as first applied for CPUs (Georgiev et al. 2015, 2017). We challenge the reader to find more examples or systems where this proportionality is not valid, as any exceptions will be very informative and improve the model.

Figure 3 suggests that the progress of nucleosynthesis follows a power law regardless of the variation of the initial metallicities and explosion energies of stars. This means that the level of complexity is more advanced in higher mass than in lower mass stars at the end of their life, even when explosion energies are taken into account, similar to other complex systems as mentioned above (Bonner 2004; Georgiev et al. 2015).

There is an additional trend that the higher the initial metallicity, the greater the increase in progress as a function of mass, with or without taking into account the explosion energies, as it can be seen in Tables 6 and 7. In our model (Georgiev et al. 2015), self-organization progresses exponentially during the lifetime of the system. Therefore, if the stellar system starts at a higher level of organization, then, it is further along its exponential trend, and, continuing that trend from a higher level, reaches much higher values of complexity within the same lifetime of the stars. This is seen visually, in the increase of the slopes of Figs. 2 and 3 with increased initial metallicity, and in the increase of the power of the corresponding fit equation in Tables 6 and 7. Those numbers can be used to calculate the exponential trends within the lifetime of stars, a subject of future work.

To reiterate, we found that the progress is in a power law dependence on the mass of the star, even when the initial metallicity and explosion energies are different. This compares to our previous studies of CPU evolution, where all of their characteristics were demonstrated to be power laws of each other (Georgiev et al. 2015, 2017). This shows that the larger the star is, the greater its progress of nucleosynthesis, as a measure for its level of complexity, similar to previous studies of size-complexity rules (Bonner 2004). By calculating the stellar lifetime, using the mass of each star, it is also possible to find the average rate of nucleosynthesis over their lifetime, including the final explosion and the action efficiency of nucleosynthesis as a function of the mass of stars. We will show this in follow up papers. The goal is to show that this power law dependence and the exponential increase during the lifetime are universal features of all complex systems, independent of their nature, be they physical, chemical, biological, technological, social, etc., and to apply it in our future work to as many other systems as possible to look for confirmations of this model and for possible exceptions.

Because in our model, the power law relations between all characteristics of complex systems are always obeyed, this may be one criterion that can be used to recognize which systems are complex. The rest, which are simpler, do not self-organize, and we cannot use any of their characteristics, to predict any other. There is a continuum of less complex and more complex systems, as some of the less complex obey certain aspects of the model, but not others. Notably, stars cannot grow in size in response of the increase of their complexity, as their mass is fixed by the primordial gas cloud. Other intermediate systems that obey only some aspects of the model are physical systems, such as Benard Cells and Vortexes, and chemical systems, such as Belousov–Zhabotinsky (BZ) reactions.

7 Conclusions

Our conclusions are not only about stellar evolution, but complex systems and complexification in general. We check our model (Georgiev et al. 2015) with a specific system—stars—to understand the process of self-organization better. Stellar evolution obeys the size-complexity rule (Bonner 2004), because the progress of nucleosynthesis as measured by the degree of grouping of nucleons is in a power law of stellar mass. Progress of nucleosynthesis confirms one aspect of our model, namely the quantity-quality transition (Georgiev et al. 2015, 2017) and the other aspects of the mutual dependence of their characteristics, such as rate of nucleosynthesis, number of nuclear additions, Free Energy Rate Density, etc., are to be explored in future papers. This gives us a new understanding of stars as complex systems. It has predictive power for progress of nucleosynthesis and the changes in the other stellar characteristics dependent on the masses of stars outside of the range included in this study, and within the lifetime of each star, which are new hypotheses to be tested. This strengthens our model and allows us to make predictions about the general behavior of other complex systems.

In this paper, we aim for generality through a very specific example. The succession of nucleosynthesis is pre-programmed in natural laws and is repeated in all stars of similar mass, as is the self-organization of all other systems. We cannot predict which two atoms will interact at any given moment, but we can predict the global statistical properties of the star, as we can predict that the number of species will increase if we increase the size of an ecosystem, or the number of different occupations will increase if a city grows in size, without being able to predict any specifics. It is in the true sense evolution—“unfolding” of something that can be predicted and is prescribed upfront towards a “future state,” in the case of stars, of heavier nuclei, in the case of ecosystems—of a larger variety of species, or in the case of cities—of increased variety of occupations and organization traits, etc. There is a small room for chance or contingency on external fluctuations, within the limits of stability of the system.

Ultimately, the existence of a complex system causes its self-organization and its self-organization causes its existence. In the example that we investigated, stellar self-organization is the natural byproduct of the star’s existence, the nucleons group and increase the complexity of the star while providing energy for its existence. On the other hand, the existence of a star is a byproduct of its self-organization—nucleosynthesis is what creates the energy to provide the outward pressure to balance gravity and maintain its existence while on the main sequence. Size-complexity examples, such as those that operate in biology, are that if cells do not differentiate, organisms cease to grow and if they do not grow, their cells cannot differentiate (Bonner 2004). This is true statistically for most of the organisms, as there are small exceptions of it, such as for slime mold. Furthermore, Carneiro shows that if new organizational units do not appear in expanding societies, they split into smaller ones, and if they do not grow in size they cannot acquire new organizational traits. There are animal species, such as alligators and turtles, and cities, such as Rome or Venice, that appear to have existed without changing, but this is only true over relatively short time-scales. If we extend the time-scale to the length of the entire evolution of that system—billions of years in biology and thousands of years in the human civilization, we will see that they always change, although at varying rates. We may even go to such lengths, as to say that civilizations that stopped self-organizing (improving, evolving), were overtaken by the ones that did, and therefore many of those ceased to exist. We can make a prediction that those trends will continue in the future as long as there are non-equilibrium thermodynamic systems of any kind, and that this happens everywhere in the Universe.

In order to show the applicability of first principles, and of the Principle of Least Action and action efficiency importance in this research, as in our model (Georgiev et al. 2015, 2017), we are planning to look for the time dependence of the rate of nucleosynthesis during the lifetime in each individual star. Based on our model, the prediction is that complexity (amount of grouping of nucleons, the progress of nucleosynthesis) increases exponentially during the star's lifetime. Another prediction is that action efficiency of nucleosynthesis will increase following a power law with the mass of stars. Whether those predictions are confirmed or rejected will either strengthen or modify our model. Another prediction based on our model is that when a gas cloud collapses into a star, there is a positive feedback between the gravity and the amount of matter in the star, therefore initially, the size of the star will increase exponentially, and then as the matter in the gas cloud is exhausted, it will reach saturation and stop growing, forming a logistic curve.

Stellar evolution is naturally connected to galactic evolution, as the metallicity of galaxies and their structure formation depends on the processes of nucleosynthesis in stars. In our future work we will explore whether galactic evolution also obeys the size-complexity rule and all other proportionality relations between the characteristics of complex systems and their exponential increase in time, to test and develop further our model.

Acknowledgments We would like to thank the Assumption University Honor's Program and the Department of Biological and Physical Sciences for financial support, as well as Assumption students Thanh Vu, Simon Trcka and Lamberto Qako for help with data collection. We would also like to thank Assumption students Nicole Bramlitt and Rebecca Gilchrist for long discussions and for proofreading the paper, and to Prof. Steven Theroux, Prof. Michael Levin, Clement Vidal and Claudio Flores Martinez for valuable feedback.

Appendix

See Tables 8, 9, 10 and 11.

Table 8 Calculated number of nucleons of each isotope when the Metallicity equals 0

$M_*(M_\odot)$	13	15	18	20	25	30	40
^1H	7.89×10^{57}	9.08×10^{57}	1.01×10^{58}	1.05×10^{58}	1.27×10^{58}	1.4×10^{58}	1.68×10^{58}
^2H	1.78×10^{41}	2.02×10^{41}	1.53×10^{41}	1.04×10^{41}	2.42×10^{41}	1.61×10^{41}	4.14×10^{41}
^3He	4.93×10^{52}	4.9×10^{52}	3.99×10^{52}	5.7×10^{52}	2.53×10^{53}	2.47×10^{53}	3.07×10^{52}
^4He	4.8×10^{57}	5.27×10^{57}	6.49×10^{57}	7.11×10^{57}	9.62×10^{57}	1.14×10^{58}	1.43×10^{58}
He_{nuc}	8.79×10^{56}	7.43×10^{56}	1.06×10^{57}	1.08×10^{57}	2.07×10^{57}	2.35×10^{57}	2.18×10^{57}
^6Li	4.37×10^{34}	1.33×10^{35}	5.23×10^{34}	4.37×10^{36}	3.22×10^{36}	1.35×10^{35}	9.03×10^{35}
^7Li	2.6×10^{47}	3.52×10^{47}	8.79×10^{46}	3.34×10^{47}	6.8×10^{48}	2.83×10^{49}	4.5×10^{46}
^9Be	2.12×10^{37}	3.86×10^{35}	1.26×10^{35}	5.38×10^{34}	1.49×10^{40}	1.51×10^{37}	6.38×10^{37}
^{10}B	3.5×10^{36}	9.94×10^{37}	4.7×10^{36}	1.88×10^{38}	3.44×10^{39}	6.2×10^{37}	2.84×10^{40}
^{11}B	3.52×10^{41}	3.95×10^{41}	8.55×10^{41}	7.83×10^{40}	1.13×10^{42}	3.92×10^{42}	3.68×10^{43}
^{12}C	8.88×10^{55}	2.06×10^{56}	2.61×10^{56}	2.53×10^{56}	3.52×10^{56}	4.04×10^{56}	5.14×10^{56}
^{13}C	1×10^{50}	7.44×10^{49}	3.15×10^{48}	1.37×10^{49}	1.76×10^{49}	1.22×10^{49}	3.84×10^{48}
^{14}N	2.19×10^{54}	2.23×10^{54}	2.26×10^{53}	6.49×10^{52}	7.08×10^{53}	1.96×10^{51}	7.05×10^{50}
^{15}N	7.64×10^{49}	8.22×10^{49}	2.87×10^{49}	1.35×10^{49}	1.4×10^{50}	2×10^{49}	7.53×10^{50}
^{16}O	5.39×10^{56}	9.26×10^{56}	1.65×10^{57}	2.53×10^{57}	3.34×10^{57}	5.76×10^{57}	1×10^{58}
^{17}O	2.02×10^{51}	1.88×10^{51}	3.34×10^{50}	8.18×10^{49}	1.78×10^{51}	2.25×10^{49}	1.7×10^{48}
^{18}O	6.94×10^{49}	5.86×10^{51}	5.55×10^{51}	3.02×10^{49}	8.08×10^{50}	2.47×10^{48}	2.55×10^{50}
^{19}F	1.4×10^{47}	2.36×10^{48}	9.47×10^{48}	1.94×10^{48}	2.05×10^{48}	1.07×10^{48}	2.85×10^{47}
^{20}Ne	1.83×10^{55}	3.92×10^{56}	5.92×10^{56}	1.09×10^{57}	6.38×10^{56}	1.02×10^{57}	3.68×10^{56}
^{21}Ne	6.49×10^{50}	4.5×10^{52}	1.09×10^{53}	5.15×10^{52}	1.59×10^{52}	6.6×10^{52}	1.29×10^{52}
^{22}Ne	2.37×10^{50}	1.93×10^{52}	3.08×10^{52}	8.29×10^{52}	2.42×10^{52}	1.03×10^{53}	8.08×10^{51}
^{23}Na	1.72×10^{53}	2.93×10^{54}	2.49×10^{54}	3.47×10^{54}	1.23×10^{54}	1.7×10^{54}	2.2×10^{53}
^{24}Mg	1.03×10^{56}	8.17×10^{55}	1.88×10^{56}	1.8×10^{56}	1.44×10^{56}	2.71×10^{56}	5.73×10^{56}
^{25}Mg	1.87×10^{53}	3.57×10^{53}	6.98×10^{53}	1.39×10^{53}	4.76×10^{52}	2.92×10^{53}	5.13×10^{53}
^{26}Mg	8.47×10^{52}	4.77×10^{53}	1.05×10^{54}	2.85×10^{53}	6×10^{52}	1.55×10^{53}	1.5×10^{53}
^{26}Al	1.19×10^{51}	1.33×10^{51}	3.99×10^{51}	5.96×10^{50}	8.98×10^{50}	3.5×10^{51}	1.66×10^{51}
^{27}Al	4.53×10^{54}	1.64×10^{54}	3.76×10^{54}	1.64×10^{54}	9.68×10^{53}	3.15×10^{54}	1.76×10^{55}
^{28}Si	9.63×10^{55}	8.77×10^{55}	1.39×10^{56}	1.19×10^{56}	4.2×10^{56}	2.97×10^{56}	1.22×10^{57}
^{29}Si	8.98×10^{53}	2.86×10^{53}	5.29×10^{53}	2.18×10^{53}	3.25×10^{53}	7.04×10^{53}	3.11×10^{54}
^{30}Si	1.7×10^{54}	1.78×10^{53}	4.13×10^{53}	1.32×10^{53}	9.03×10^{52}	3.05×10^{53}	4.86×10^{54}
^{31}P	5.85×10^{53}	6.74×10^{52}	1.58×10^{53}	9.59×10^{52}	1.01×10^{53}	1.4×10^{53}	1.92×10^{54}
^{32}S	2.84×10^{55}	3.83×10^{55}	4.87×10^{55}	6.36×10^{55}	2.22×10^{56}	1.39×10^{56}	4.47×10^{56}
^{33}S	1.08×10^{53}	9.04×10^{52}	1.23×10^{53}	2.37×10^{53}	3.28×10^{53}	1.98×10^{53}	9.7×10^{53}
^{34}S	3.34×10^{53}	2.42×10^{53}	3.41×10^{53}	5.87×10^{53}	5.08×10^{53}	1.01×10^{53}	1.9×10^{54}
^{36}S	1.77×10^{49}	1.71×10^{48}	6.4×10^{48}	3.07×10^{48}	4.08×10^{47}	8.43×10^{47}	3.82×10^{49}
^{35}Cl	6.56×10^{52}	1.75×10^{52}	3.16×10^{52}	8.24×10^{52}	6.49×10^{52}	2.75×10^{52}	2.59×10^{53}
^{37}Cl	3.64×10^{51}	6.98×10^{51}	1.09×10^{52}	4.62×10^{52}	7.32×10^{52}	1.84×10^{52}	1.16×10^{53}
^{36}Ar	3.88×10^{54}	6.32×10^{54}	6.79×10^{54}	1.16×10^{55}	3.71×10^{55}	2.35×10^{55}	5.83×10^{55}
^{38}Ar	6.26×10^{52}	7.46×10^{52}	2.04×10^{53}	4.62×10^{53}	4.3×10^{53}	4.17×10^{52}	1.33×10^{54}
^{40}Ar	9.59×10^{46}	2.13×10^{46}	4.74×10^{46}	1.27×10^{47}	2.12×10^{46}	5.58×10^{45}	1.55×10^{47}

(continued)

Table 8 (continued)

$M_*(M_\odot)$	13	15	18	20	25	30	40
^{39}K	6.01×10^{51}	9.41×10^{51}	2.31×10^{52}	5.27×10^{52}	7.55×10^{52}	1.56×10^{52}	1.41×10^{53}
^{40}K	1.37×10^{48}	1.09×10^{48}	2.35×10^{48}	1.56×10^{49}	9.63×10^{48}	1.19×10^{48}	1.25×10^{49}
^{41}K	4.24×10^{50}	9.64×10^{50}	2.06×10^{51}	1.13×10^{52}	2.49×10^{52}	4.06×10^{51}	3.35×10^{52}
^{40}Ca	3.5×10^{54}	5.28×10^{54}	5.27×10^{54}	7.45×10^{54}	2.97×10^{55}	2.08×10^{55}	4.47×10^{55}
^{42}Ca	1.17×10^{51}	1.47×10^{51}	4.34×10^{51}	1.53×10^{52}	8.94×10^{51}	1.03×10^{51}	2.6×10^{52}
^{43}Ca	7.74×10^{49}	5.91×10^{49}	4.06×10^{49}	8.9×10^{49}	1.89×10^{49}	2.31×10^{48}	1.16×10^{49}
^{44}Ca	2.01×10^{52}	2.65×10^{52}	1.74×10^{52}	1.74×10^{52}	1.17×10^{52}	6.52×10^{51}	1.04×10^{52}
^{46}Ca	1.28×10^{45}	2.11×10^{45}	1.11×10^{46}	1.56×10^{47}	3.33×10^{46}	7.25×10^{44}	6.23×10^{45}
^{48}Ca	1.86×10^{40}	5.04×10^{43}	4.93×10^{41}	5.27×10^{41}	1.51×10^{46}	4.25×10^{41}	1.56×10^{40}
^{45}Sc	2.54×10^{49}	4.83×10^{49}	6.64×10^{49}	3.05×10^{50}	7.38×10^{50}	1.81×10^{50}	6.96×10^{50}
^{46}Ti	7.52×10^{51}	3.25×10^{51}	4.86×10^{51}	7.16×10^{51}	3.81×10^{51}	6.32×10^{50}	1.29×10^{52}
^{47}Ti	1.05×10^{52}	4.62×10^{51}	6.35×10^{51}	5.1×10^{51}	7.07×10^{49}	4.55×10^{49}	1.11×10^{50}
^{48}Ti	7.59×10^{52}	1×10^{53}	9.16×10^{52}	1.06×10^{53}	1.86×10^{53}	2.16×10^{53}	2.93×10^{53}
^{49}Ti	2.72×10^{51}	3.98×10^{51}	3.59×10^{51}	4.72×10^{51}	8.43×10^{51}	1.06×10^{52}	1.44×10^{52}
^{50}Ti	1.41×10^{45}	1.14×10^{45}	2.06×10^{45}	2.19×10^{45}	2.68×10^{45}	5.5×10^{43}	1.13×10^{46}
^{50}V	1.62×10^{46}	1.26×10^{46}	5.51×10^{46}	1.63×10^{47}	1.4×10^{47}	3.16×10^{45}	7.82×10^{47}
^{51}V	1.98×10^{52}	1.29×10^{52}	1.49×10^{52}	1.32×10^{52}	1.08×10^{52}	1.27×10^{52}	2.11×10^{52}
^{50}Cr	1.25×10^{52}	1.86×10^{52}	3.16×10^{52}	2.66×10^{52}	5.85×10^{52}	4.58×10^{52}	2.13×10^{53}
^{52}Cr	1.05×10^{54}	1.31×10^{54}	1.35×10^{54}	1.63×10^{54}	3.32×10^{54}	3.74×10^{54}	4.73×10^{54}
^{53}Cr	5.94×10^{52}	8.1×10^{52}	7.7×10^{52}	9.92×10^{52}	1.8×10^{53}	2.18×10^{53}	3.13×10^{53}
^{54}Cr	2.81×10^{47}	4.95×10^{47}	3.89×10^{48}	3.81×10^{48}	1.12×10^{49}	2.06×10^{47}	1.03×10^{50}
^{55}Mn	1.59×10^{53}	2.22×10^{53}	2.08×10^{53}	2.71×10^{53}	5.15×10^{53}	6.18×10^{53}	8.56×10^{53}
^{54}Fe	8.73×10^{53}	1.49×10^{54}	1.68×10^{54}	1.7×10^{54}	3.96×10^{54}	4.9×10^{54}	1.11×10^{55}
^{56}Fe	8.38×10^{55}	8.38×10^{55}	8.38×10^{55}	8.38×10^{55}	8.38×10^{55}	8.38×10^{55}	8.41×10^{55}
^{57}Fe	1.19×10^{54}	1.37×10^{54}	1.05×10^{54}	1.04×10^{54}	5.59×10^{53}	5.79×10^{53}	6×10^{53}
^{58}Fe	6.79×10^{46}	2.11×10^{47}	9.67×10^{47}	1.26×10^{48}	2.77×10^{48}	2.71×10^{47}	1.69×10^{49}
^{59}Co	2.11×10^{53}	1.58×10^{53}	1.93×10^{53}	1.8×10^{53}	1.88×10^{52}	2.96×10^{51}	2.24×10^{51}
^{58}Ni	4.61×10^{53}	4.96×10^{53}	4.59×10^{53}	4.53×10^{53}	3.51×10^{53}	4.58×10^{53}	6.88×10^{53}
^{60}Ni	2.54×10^{54}	1.94×10^{54}	1.88×10^{54}	1.61×10^{54}	1.77×10^{53}	4.31×10^{51}	5.25×10^{51}
^{61}Ni	4.32×10^{52}	3.76×10^{52}	2.54×10^{52}	2.22×10^{52}	6.91×10^{50}	7.23×10^{48}	1.58×10^{48}
^{62}Ni	2.32×10^{52}	1.81×10^{52}	1.62×10^{52}	1.43×10^{52}	4.65×10^{50}	1.96×10^{48}	6.42×10^{47}
^{64}Ni	4.59×10^{42}	1.19×10^{44}	1.31×10^{43}	7.31×10^{43}	3.92×10^{45}	3.9×10^{43}	9.23×10^{42}
^{63}Cu	5.86×10^{51}	4.24×10^{51}	4.7×10^{51}	4.12×10^{51}	1.47×10^{50}	2.84×10^{47}	2.47×10^{46}
^{65}Cu	2.56×10^{50}	2.9×10^{50}	1.95×10^{50}	1.7×10^{50}	9.33×10^{48}	1.23×10^{45}	1.56×10^{44}
^{64}Zn	1.5×10^{53}	1.46×10^{53}	1.14×10^{53}	9.93×10^{52}	3.04×10^{51}	3.62×10^{47}	5.22×10^{46}
^{66}Zn	8.19×10^{50}	1.28×10^{51}	6.05×10^{50}	5.08×10^{50}	1.9×10^{49}	4.23×10^{45}	6.16×10^{44}
^{67}Zn	1.94×10^{49}	2.79×10^{49}	1.56×10^{49}	1.26×10^{49}	2.59×10^{47}	7.56×10^{44}	1.01×10^{44}
^{68}Zn	3.52×10^{49}	3.88×10^{49}	4.82×10^{49}	4.11×10^{49}	9.64×10^{47}	1.56×10^{45}	3.14×10^{44}
^{70}Zn	8.31×10^{41}	4.3×10^{43}	7.89×10^{42}	3.01×10^{43}	4.76×10^{43}	2.23×10^{43}	5.33×10^{41}
^{69}Ga	9.35×10^{48}	6.73×10^{48}	7.31×10^{48}	5.91×10^{48}	1.05×10^{47}	6.06×10^{44}	5.1×10^{42}
^{71}Ga	1.02×10^{43}	1.34×10^{44}	2.2×10^{43}	1.1×10^{44}	2.68×10^{44}	2.8×10^{43}	1.63×10^{42}
$^{140}\chi$	0	0	0	0	0	0	0
$^{100}\rho$	0	0	0	0	0	0	0

Table 9 Number of nucleons of each isotope when the Metallicity equals 0.001

$M_*(M_\odot)$	13	15	18	20	25	30	40
^1H	7.71×10^{57}	8.92×10^{57}	1.01×10^{58}	1.01×10^{58}	1.17×10^{58}	1.32×10^{58}	1.55×10^{58}
^2H	8.61×10^{42}	4.05×10^{45}	3.68×10^{42}	5.21×10^{41}	5.34×10^{41}	6.02×10^{41}	8.42×10^{43}
^3He	1.71×10^{53}	1.83×10^{53}	1.88×10^{53}	1.92×10^{53}	1.51×10^{53}	1.72×10^{53}	1.45×10^{53}
^4He	4.62×10^{57}	6.18×10^{57}	7.83×10^{57}	7.11×10^{57}	8.35×10^{57}	1.00×10^{58}	1.31×10^{58}
He_{nuc}	7.00×10^{57}	1.65×10^{57}	2.04×10^{57}	1.08×10^{57}	8.03×10^{56}	9.82×10^{56}	9.82×10^{56}
^6Li	2.84×10^{40}	1.33×10^{40}	1.06×10^{40}	2.87×10^{34}	2.81×10^{36}	3.29×10^{34}	2.74×10^{41}
^7Li	7.59×10^{47}	4.50×10^{44}	6.98×10^{44}	1.01×10^{47}	1.09×10^{45}	1.34×10^{45}	8.00×10^{45}
^9Be	2.79×10^{40}	1.09×10^{42}	1.12×10^{41}	1.71×10^{34}	2.46×10^{32}	0.00×10^{00}	4.86×10^{41}
^{10}B	1.38×10^{46}	7.20×10^{45}	7.87×10^{45}	2.84×10^{45}	9.10×10^{45}	2.69×10^{45}	2.41×10^{45}
^{11}B	6.12×10^{46}	3.21×10^{46}	3.52×10^{46}	1.23×10^{46}	4.10×10^{46}	1.21×10^{46}	8.40×10^{45}
^{12}C	1.28×10^{56}	1.02×10^{56}	1.55×10^{56}	1.53×10^{56}	2.58×10^{56}	1.45×10^{56}	8.83×10^{55}
^{13}C	2.00×10^{53}	6.44×10^{52}	8.62×10^{52}	2.35×10^{52}	1.18×10^{53}	9.82×10^{52}	3.38×10^{53}
^{14}N	1.09×10^{55}	4.29×10^{54}	5.35×10^{54}	1.55×10^{55}	1.10×10^{55}	7.41×10^{54}	1.04×10^{55}
^{15}N	9.07×10^{51}	1.03×10^{51}	1.10×10^{51}	1.64×10^{51}	8.67×10^{51}	4.54×10^{50}	2.92×10^{51}
^{16}O	6.04×10^{56}	3.52×10^{56}	5.05×10^{56}	2.61×10^{57}	4.58×10^{57}	6.38×10^{57}	1.00×10^{58}
^{17}O	8.34×10^{52}	3.07×10^{52}	2.90×10^{52}	2.61×10^{52}	3.34×10^{52}	5.83×10^{52}	3.43×10^{52}
^{18}O	2.17×10^{54}	4.38×10^{53}	3.67×10^{53}	9.69×10^{51}	8.44×10^{52}	3.22×10^{52}	3.15×10^{52}
^{19}F	3.68×10^{51}	2.37×10^{50}	3.98×10^{50}	3.35×10^{51}	7.33×10^{50}	2.86×10^{51}	5.35×10^{51}
^{20}Ne	7.91×10^{55}	2.28×10^{56}	2.12×10^{56}	7.51×10^{56}	1.46×10^{57}	1.74×10^{57}	3.44×10^{56}
^{21}Ne	2.22×10^{53}	8.08×10^{52}	1.05×10^{53}	1.64×10^{53}	6.10×10^{53}	7.32×10^{53}	1.33×10^{53}
^{22}Ne	1.59×10^{54}	3.28×10^{53}	5.69×10^{53}	1.40×10^{54}	1.76×10^{54}	1.87×10^{54}	1.05×10^{54}
^{23}Na	6.48×10^{53}	2.35×10^{54}	2.50×10^{54}	2.17×10^{54}	9.69×10^{54}	8.22×10^{54}	1.08×10^{54}
^{24}Mg	7.62×10^{55}	7.63×10^{55}	7.10×10^{55}	2.90×10^{56}	2.14×10^{56}	3.43×10^{56}	8.43×10^{56}
^{25}Mg	1.68×10^{54}	1.05×10^{54}	1.13×10^{54}	2.87×10^{54}	2.08×10^{54}	4.25×10^{54}	2.65×10^{54}
^{26}Mg	9.87×10^{53}	1.37×10^{54}	1.11×10^{54}	2.91×10^{54}	2.38×10^{54}	5.10×10^{54}	1.32×10^{54}
^{26}Al	1.21×10^{52}	2.19×10^{51}	3.73×10^{51}	4.28×10^{51}	3.33×10^{51}	9.17×10^{51}	5.99×10^{51}
^{27}Al	4.28×10^{54}	2.81×10^{54}	2.77×10^{54}	8.36×10^{54}	6.05×10^{54}	1.05×10^{55}	3.61×10^{55}
^{28}Si	1.08×10^{56}	5.14×10^{55}	1.83×10^{56}	1.53×10^{56}	1.44×10^{56}	1.98×10^{56}	1.06×10^{57}
^{29}Si	1.70×10^{54}	4.58×10^{53}	7.20×10^{53}	1.33×10^{54}	5.27×10^{53}	1.23×10^{54}	7.23×10^{54}
^{30}Si	2.22×10^{54}	5.09×10^{53}	6.40×10^{53}	9.29×10^{53}	3.29×10^{53}	8.49×10^{53}	1.21×10^{55}
^{31}P	6.36×10^{53}	9.73×10^{52}	2.24×10^{53}	2.36×10^{53}	1.28×10^{53}	2.44×10^{53}	4.17×10^{54}
^{32}S	4.43×10^{55}	1.96×10^{55}	9.44×10^{55}	6.73×10^{55}	6.60×10^{55}	9.34×10^{55}	3.94×10^{56}
^{33}S	2.35×10^{53}	5.70×10^{52}	3.61×10^{53}	1.19×10^{53}	9.73×10^{52}	1.66×10^{53}	1.00×10^{54}
^{34}S	1.10×10^{54}	2.86×10^{53}	7.63×10^{53}	3.71×10^{53}	2.14×10^{53}	4.41×10^{53}	2.54×10^{54}
^{36}S	8.97×10^{50}	1.86×10^{50}	3.83×10^{50}	4.85×10^{50}	7.82×10^{50}	1.08×10^{51}	6.66×10^{50}
^{35}Cl	9.77×10^{52}	9.70×10^{51}	9.77×10^{52}	2.79×10^{52}	2.52×10^{52}	4.01×10^{52}	4.22×10^{53}
^{37}Cl	1.72×10^{52}	3.15×10^{51}	4.95×10^{52}	1.22×10^{52}	1.55×10^{52}	2.30×10^{52}	7.56×10^{52}
^{36}Ar	6.82×10^{54}	2.95×10^{54}	1.46×10^{55}	1.21×10^{55}	1.11×10^{55}	1.62×10^{55}	5.46×10^{55}
^{38}Ar	3.21×10^{53}	2.71×10^{52}	4.93×10^{53}	1.28×10^{53}	9.88×10^{52}	1.95×10^{53}	1.00×10^{54}
^{40}Ar	2.11×10^{50}	5.87×10^{49}	1.23×10^{50}	6.50×10^{49}	1.18×10^{50}	1.41×10^{50}	8.18×10^{49}
^{39}K	3.01×10^{52}	4.97×10^{51}	6.82×10^{52}	1.66×10^{52}	1.82×10^{52}	2.84×10^{52}	1.33×10^{53}
^{40}K	3.19×10^{49}	2.34×10^{48}	4.70×10^{49}	9.07×10^{48}	1.19×10^{49}	1.14×10^{49}	2.85×10^{49}

(continued)

Table 9 (continued)

$M_*(M_\odot)$	13	15	18	20	25	30	40
^{41}K	2.89×10^{51}	3.75×10^{50}	1.31×10^{52}	2.48×10^{51}	2.66×10^{51}	4.77×10^{51}	1.82×10^{52}
^{40}Ca	5.67×10^{54}	2.07×10^{54}	9.65×10^{54}	1.10×10^{55}	9.51×10^{54}	1.40×10^{55}	4.38×10^{55}
^{42}Ca	7.93×10^{51}	4.85×10^{50}	1.62×10^{52}	3.34×10^{51}	2.98×10^{51}	5.56×10^{51}	2.74×10^{52}
^{43}Ca	4.97×10^{50}	4.01×10^{50}	2.93×10^{50}	2.29×10^{50}	3.47×10^{50}	4.25×10^{50}	3.37×10^{50}
^{44}Ca	2.97×10^{52}	2.48×10^{52}	2.24×10^{52}	5.85×10^{51}	9.53×10^{51}	7.71×10^{51}	1.32×10^{52}
^{46}Ca	7.97×10^{49}	1.69×10^{49}	3.05×10^{49}	4.01×10^{49}	3.62×10^{49}	8.49×10^{49}	1.46×10^{50}
^{48}Ca	3.78×10^{50}	1.16×10^{50}	1.46×10^{50}	1.50×10^{50}	1.88×10^{50}	2.23×10^{50}	2.60×10^{50}
^{45}Sc	3.02×10^{50}	6.08×10^{49}	4.25×10^{50}	2.59×10^{50}	3.88×10^{50}	5.05×10^{50}	1.19×10^{51}
^{46}Ti	4.77×10^{51}	1.49×10^{52}	6.12×10^{51}	1.63×10^{51}	1.57×10^{51}	2.58×10^{51}	1.20×10^{52}
^{47}Ti	3.05×10^{51}	1.53×10^{52}	1.80×10^{51}	5.15×10^{50}	4.02×10^{50}	5.39×10^{50}	7.61×10^{50}
^{48}Ti	1.16×10^{53}	6.22×10^{52}	1.20×10^{53}	1.64×10^{53}	1.40×10^{53}	2.00×10^{53}	3.03×10^{53}
^{49}Ti	4.61×10^{51}	2.32×10^{51}	4.95×10^{51}	8.12×10^{51}	6.80×10^{51}	1.01×10^{52}	1.68×10^{52}
^{50}Ti	7.57×10^{50}	1.94×10^{50}	2.23×10^{50}	6.05×10^{50}	1.02×10^{51}	1.37×10^{51}	1.39×10^{51}
^{50}V	1.32×10^{49}	1.34×10^{48}	2.17×10^{48}	1.02×10^{49}	1.57×10^{49}	2.47×10^{49}	1.11×10^{50}
^{51}V	1.07×10^{52}	1.94×10^{52}	9.06×10^{51}	9.86×10^{51}	9.01×10^{51}	1.19×10^{52}	2.32×10^{52}
^{50}Cr	2.52×10^{52}	1.59×10^{52}	3.58×10^{52}	2.84×10^{52}	3.29×10^{52}	4.20×10^{52}	1.62×10^{53}
^{52}Cr	1.44×10^{54}	3.63×10^{53}	1.82×10^{54}	2.66×10^{54}	2.61×10^{54}	3.50×10^{54}	4.72×10^{54}
^{53}Cr	8.90×10^{52}	3.38×10^{52}	1.09×10^{53}	1.64×10^{53}	1.33×10^{53}	2.12×10^{53}	3.07×10^{53}
^{54}Cr	2.24×10^{51}	5.75×10^{50}	7.14×10^{50}	1.69×10^{51}	2.61×10^{51}	3.28×10^{51}	3.40×10^{51}
^{55}Mn	2.72×10^{53}	9.59×10^{52}	3.37×10^{53}	4.55×10^{53}	3.58×10^{53}	6.00×10^{53}	9.41×10^{53}
^{54}Fe	1.83×10^{54}	8.47×10^{53}	2.47×10^{54}	2.84×10^{54}	2.69×10^{54}	3.96×10^{54}	1.25×10^{55}
^{56}Fe	8.70×10^{55}	8.48×10^{55}	8.52×10^{55}	8.49×10^{55}	8.52×10^{55}	8.53×10^{55}	8.56×10^{55}
^{57}Fe	1.69×10^{54}	2.10×10^{54}	1.52×10^{54}	8.64×10^{53}	6.35×10^{53}	6.98×10^{53}	6.89×10^{53}
^{58}Fe	6.80×10^{52}	1.57×10^{52}	1.86×10^{52}	5.98×10^{52}	9.32×10^{52}	1.11×10^{53}	1.22×10^{53}
^{59}Co	1.07×10^{53}	2.78×10^{53}	7.55×10^{52}	7.44×10^{52}	4.10×10^{52}	6.46×10^{52}	5.29×10^{52}
^{58}Ni	6.10×10^{53}	1.09×10^{54}	2.11×10^{54}	4.24×10^{53}	3.01×10^{53}	4.54×10^{53}	9.34×10^{53}
^{60}Ni	1.94×10^{54}	3.08×10^{54}	1.40×10^{54}	3.53×10^{53}	7.87×10^{52}	2.32×10^{53}	1.21×10^{53}
^{61}Ni	6.05×10^{52}	8.29×10^{52}	5.55×10^{52}	1.46×10^{52}	1.88×10^{52}	2.52×10^{52}	2.47×10^{52}
^{62}Ni	1.21×10^{53}	2.24×10^{53}	3.56×10^{53}	3.03×10^{52}	4.44×10^{52}	6.17×10^{52}	9.10×10^{52}
^{64}Ni	1.94×10^{52}	2.19×10^{51}	2.75×10^{51}	2.84×10^{52}	5.17×10^{52}	7.74×10^{52}	9.38×10^{52}
^{63}Cu	9.64×10^{51}	6.55×10^{51}	3.33×10^{51}	1.15×10^{52}	1.72×10^{52}	2.36×10^{52}	1.68×10^{52}
^{65}Cu	5.81×10^{51}	1.06×10^{51}	1.37×10^{51}	9.64×10^{51}	1.78×10^{52}	2.67×10^{52}	3.47×10^{52}
^{64}Zn	8.56×10^{52}	7.79×10^{52}	6.37×10^{52}	1.40×10^{52}	1.08×10^{52}	1.93×10^{52}	1.55×10^{52}
^{66}Zn	1.18×10^{52}	4.74×10^{51}	9.26×10^{51}	1.34×10^{52}	2.31×10^{52}	3.70×10^{52}	5.73×10^{52}
^{67}Zn	1.25×10^{51}	1.69×10^{50}	2.24×10^{50}	2.44×10^{51}	4.62×10^{51}	7.26×10^{51}	5.01×10^{51}
^{68}Zn	8.23×10^{51}	6.97×10^{50}	9.41×10^{50}	1.18×10^{52}	2.23×10^{52}	3.77×10^{52}	5.55×10^{52}
^{70}Zn	2.74×10^{50}	1.26×10^{49}	2.10×10^{49}	3.43×10^{49}	2.81×10^{49}	8.83×10^{49}	1.66×10^{50}
^{69}Ga	8.52×10^{50}	8.90×10^{49}	1.35×10^{50}	1.49×10^{51}	2.79×10^{51}	4.47×10^{51}	5.67×10^{51}
^{71}Ga	7.59×10^{50}	6.14×10^{49}	9.22×10^{49}	1.31×10^{51}	2.41×10^{51}	4.60×10^{51}	6.53×10^{51}
$^{140}\chi$	4.50×10^{57}	5.70×10^{57}	7.16×10^{57}	7.14×10^{57}	8.83×10^{57}	1.09×10^{58}	1.54×10^{58}
$^{100}\rho$	1.56×10^{55}	1.80×10^{55}	2.16×10^{55}	2.40×10^{55}	2.99×10^{55}	3.59×10^{55}	4.79×10^{55}

Table 10 Number of nucleons of each isotope of each star when the Metallicity equals 0.004

$M_*(M_\odot)$	13	15	18	20	25	30	40
^1H	7.63×10^{57}	8.52×10^{57}	8.95×10^{57}	1.07×10^{58}	1.22×10^{58}	1.21×10^{58}	1.23×10^{58}
^2H	1.28×10^{43}	2.38×10^{43}	5.47×10^{43}	1.05×10^{44}	3.87×10^{41}	8.30×10^{43}	3.16×10^{41}
^3He	2.04×10^{53}	1.90×10^{53}	2.68×10^{53}	2.10×10^{53}	2.22×10^{53}	2.20×10^{53}	2.16×10^{53}
^4He	4.84×10^{57}	5.93×10^{57}	7.26×10^{57}	8.42×10^{57}	1.02×10^{58}	9.49×10^{57}	9.73×10^{57}
He_{nuc}	9.15×10^{56}	1.40×10^{57}	1.83×10^{57}	2.38×10^{57}	2.61×10^{57}	4.31×10^{56}	-2.3×10^{57}
^6Li	4.12×10^{40}	7.79×10^{40}	1.80×10^{41}	3.46×10^{41}	6.22×10^{34}	2.67×10^{41}	2.56×10^{37}
^7Li	3.77×10^{42}	8.62×10^{43}	6.04×10^{44}	1.87×10^{45}	1.22×10^{44}	9.59×10^{44}	5.64×10^{45}
^9Be	1.28×10^{41}	1.29×10^{41}	1.89×10^{41}	6.60×10^{41}	5.85×10^{33}	4.53×10^{41}	7.19×10^{35}
^{10}B	3.05×10^{46}	3.05×10^{46}	1.86×10^{47}	3.49×10^{46}	3.70×10^{46}	3.64×10^{46}	7.87×10^{45}
^{11}B	1.38×10^{47}	1.38×10^{47}	8.35×10^{47}	5.51×10^{49}	1.66×10^{47}	1.64×10^{47}	3.35×10^{46}
^{12}C	1.05×10^{56}	1.06×10^{56}	1.26×10^{56}	1.17×10^{56}	1.58×10^{56}	2.18×10^{56}	5.49×10^{56}
^{13}C	2.25×10^{53}	2.50×10^{53}	7.22×10^{55}	3.50×10^{53}	4.58×10^{53}	4.06×10^{53}	4.41×10^{53}
^{14}N	1.09×10^{55}	1.55×10^{55}	8.71×10^{55}	2.20×10^{55}	3.77×10^{55}	2.40×10^{55}	3.11×10^{55}
^{15}N	8.19×10^{51}	1.04×10^{52}	6.41×10^{55}	3.33×10^{52}	1.14×10^{53}	5.96×10^{51}	6.00×10^{51}
^{16}O	4.61×10^{56}	3.50×10^{56}	6.24×10^{56}	1.19×10^{57}	2.64×10^{57}	5.74×10^{57}	9.53×10^{57}
^{17}O	1.05×10^{53}	1.04×10^{53}	1.33×10^{54}	1.25×10^{53}	1.34×10^{53}	1.44×10^{53}	1.90×10^{53}
^{18}O	2.34×10^{54}	1.16×10^{54}	6.23×10^{55}	1.53×10^{54}	1.02×10^{54}	5.21×10^{52}	1.00×10^{54}
^{19}F	2.37×10^{51}	2.55×10^{51}	1.45×10^{52}	6.42×10^{51}	1.02×10^{53}	1.39×10^{52}	1.25×10^{51}
^{20}Ne	1.58×10^{56}	1.49×10^{56}	2.40×10^{56}	3.32×10^{56}	9.82×10^{56}	1.12×10^{57}	2.25×10^{57}
^{21}Ne	2.22×10^{53}	1.71×10^{53}	1.20×10^{54}	3.09×10^{53}	4.59×10^{53}	8.64×10^{53}	1.40×10^{54}
^{22}Ne	1.26×10^{54}	9.03×10^{53}	4.99×10^{54}	2.59×10^{54}	5.47×10^{54}	7.40×10^{54}	3.44×10^{54}
^{23}Na	1.80×10^{54}	9.81×10^{53}	7.94×10^{54}	4.87×10^{54}	7.49×10^{54}	1.69×10^{55}	3.13×10^{55}
^{24}Mg	5.25×10^{55}	8.94×10^{55}	8.30×10^{55}	1.16×10^{56}	2.79×10^{56}	2.61×10^{56}	4.54×10^{56}
^{25}Mg	1.65×10^{54}	2.34×10^{54}	9.81×10^{54}	2.64×10^{54}	7.23×10^{54}	8.49×10^{54}	1.44×10^{55}
^{26}Mg	1.49×10^{54}	2.31×10^{54}	7.71×10^{54}	2.25×10^{54}	8.04×10^{54}	8.47×10^{54}	1.59×10^{55}
^{26}Al	3.21×10^{51}	5.74×10^{51}	2.71×10^{52}	1.04×10^{52}	4.42×10^{51}	1.41×10^{52}	2.67×10^{52}
^{27}Al	2.65×10^{54}	3.96×10^{54}	7.53×10^{54}	6.14×10^{54}	1.34×10^{55}	2.02×10^{55}	3.64×10^{55}
^{28}Si	7.32×10^{55}	1.23×10^{56}	1.13×10^{56}	1.49×10^{56}	1.43×10^{56}	4.73×10^{56}	6.26×10^{56}
^{29}Si	6.48×10^{53}	1.34×10^{54}	3.23×10^{54}	1.52×10^{54}	2.31×10^{54}	4.04×10^{54}	5.45×10^{54}
^{30}Si	7.86×10^{53}	1.55×10^{54}	4.66×10^{54}	1.93×10^{54}	1.89×10^{54}	5.77×10^{54}	7.79×10^{54}
^{31}P	1.80×10^{53}	3.28×10^{53}	8.82×10^{53}	4.64×10^{53}	4.62×10^{53}	1.22×10^{54}	1.77×10^{54}
^{32}S	3.21×10^{55}	4.12×10^{55}	4.87×10^{55}	6.17×10^{55}	4.17×10^{55}	2.28×10^{56}	2.71×10^{56}
^{33}S	1.09×10^{53}	2.19×10^{53}	3.14×10^{53}	2.26×10^{53}	1.99×10^{53}	5.20×10^{53}	5.45×10^{53}
^{34}S	5.11×10^{53}	1.15×10^{54}	2.85×10^{54}	1.09×10^{54}	8.65×10^{53}	2.43×10^{54}	2.81×10^{54}
^{36}S	8.50×10^{50}	1.19×10^{51}	9.33×10^{51}	1.63×10^{51}	3.21×10^{51}	7.31×10^{51}	1.43×10^{52}
^{35}Cl	3.19×10^{52}	6.24×10^{52}	1.31×10^{53}	6.54×10^{52}	6.35×10^{52}	1.47×10^{53}	1.87×10^{53}
^{37}Cl	1.21×10^{52}	1.98×10^{52}	5.51×10^{52}	2.46×10^{52}	3.63×10^{52}	9.33×10^{52}	1.35×10^{53}
^{36}Ar	5.39×10^{54}	5.16×10^{54}	8.34×10^{54}	9.57×10^{54}	5.67×10^{54}	3.70×10^{55}	4.17×10^{55}
^{38}Ar	1.99×10^{53}	5.92×10^{53}	8.90×10^{53}	3.67×10^{53}	3.20×10^{53}	1.17×10^{54}	1.37×10^{54}
^{40}Ar	2.75×10^{50}	4.48×10^{50}	3.05×10^{51}	3.95×10^{50}	4.28×10^{50}	1.11×10^{51}	1.38×10^{51}
^{39}K	2.13×10^{52}	3.83×10^{52}	8.62×10^{52}	3.74×10^{52}	3.16×10^{52}	1.22×10^{53}	1.19×10^{53}
^{40}K	9.32×10^{48}	1.09×10^{49}	8.00×10^{49}	2.22×10^{49}	4.00×10^{49}	9.47×10^{49}	8.88×10^{49}

(continued)

Table 10 (continued)

$M_*(M_\odot)$	13	15	18	20	25	30	40
^{41}K	2.24×10^{51}	2.95×10^{51}	8.52×10^{51}	3.75×10^{51}	4.11×10^{51}	1.56×10^{52}	1.62×10^{52}
^{40}Ca	4.68×10^{54}	3.69×10^{54}	7.33×10^{54}	7.79×10^{54}	4.52×10^{54}	3.10×10^{55}	3.39×10^{55}
^{42}Ca	4.98×10^{51}	1.15×10^{52}	2.17×10^{52}	8.76×10^{51}	7.19×10^{51}	3.17×10^{52}	3.57×10^{52}
^{43}Ca	4.08×10^{50}	4.31×10^{50}	2.31×10^{51}	6.26×10^{50}	9.52×10^{50}	1.18×10^{51}	1.71×10^{51}
^{44}Ca	2.48×10^{52}	2.75×10^{52}	4.07×10^{52}	2.50×10^{52}	3.17×10^{52}	1.81×10^{52}	2.25×10^{52}
^{46}Ca	8.83×10^{49}	1.41×10^{50}	8.16×10^{50}	1.32×10^{50}	1.53×10^{50}	2.36×10^{50}	3.88×10^{50}
^{48}Ca	3.86×10^{50}	4.82×10^{50}	2.84×10^{51}	6.01×10^{50}	7.41×10^{50}	7.76×10^{50}	9.34×10^{50}
^{45}Sc	2.59×10^{50}	2.97×10^{50}	1.62×10^{51}	4.32×10^{50}	6.35×10^{50}	1.31×10^{51}	1.96×10^{51}
^{46}Ti	6.18×10^{51}	7.29×10^{51}	9.81×10^{51}	4.24×10^{51}	6.16×10^{51}	1.46×10^{52}	1.68×10^{52}
^{47}Ti	6.47×10^{51}	4.07×10^{51}	5.57×10^{51}	1.59×10^{51}	7.74×10^{51}	1.75×10^{51}	2.44×10^{51}
^{48}Ti	9.79×10^{52}	8.88×10^{52}	1.64×10^{53}	1.37×10^{53}	1.07×10^{53}	2.79×10^{53}	3.21×10^{53}
^{49}Ti	4.18×10^{51}	3.37×10^{51}	8.68×10^{51}	6.22×10^{51}	4.90×10^{51}	1.63×10^{52}	2.06×10^{52}
^{50}Ti	7.01×10^{50}	6.72×10^{50}	4.46×10^{51}	1.29×10^{51}	2.69×10^{51}	4.91×10^{51}	9.22×10^{51}
^{50}V	5.52×10^{48}	5.92×10^{48}	4.86×10^{49}	1.21×10^{49}	2.37×10^{49}	6.44×10^{49}	1.16×10^{50}
^{51}V	1.57×10^{52}	1.15×10^{52}	1.47×10^{52}	9.59×10^{51}	1.69×10^{52}	2.12×10^{52}	2.64×10^{52}
^{50}Cr	2.10×10^{52}	3.25×10^{52}	4.00×10^{52}	3.67×10^{52}	2.96×10^{52}	1.03×10^{53}	1.45×10^{53}
^{52}Cr	1.33×10^{54}	1.15×10^{54}	2.17×10^{54}	1.99×10^{54}	1.49×10^{54}	4.47×10^{54}	4.87×10^{54}
^{53}Cr	8.12×10^{52}	7.07×10^{52}	1.47×10^{53}	1.25×10^{53}	8.64×10^{52}	2.91×10^{53}	3.31×10^{53}
^{54}Cr	2.10×10^{51}	2.20×10^{51}	1.28×10^{52}	3.73×10^{51}	6.94×10^{51}	1.04×10^{52}	1.66×10^{52}
^{55}Mn	2.40×10^{53}	2.05×10^{53}	5.61×10^{53}	3.94×10^{53}	2.80×10^{53}	9.12×10^{53}	1.00×10^{54}
^{54}Fe	1.49×10^{54}	1.47×10^{54}	3.15×10^{54}	2.84×10^{54}	1.69×10^{54}	8.49×10^{54}	1.12×10^{55}
^{56}Fe	8.70×10^{55}	8.74×10^{55}	1.04×10^{56}	8.86×10^{55}	8.95×10^{55}	8.94×10^{55}	8.95×10^{55}
^{57}Fe	1.26×10^{54}	1.35×10^{54}	4.79×10^{54}	1.46×10^{54}	1.26×10^{54}	8.56×10^{53}	8.76×10^{53}
^{58}Fe	5.71×10^{52}	5.86×10^{52}	3.19×10^{53}	1.10×10^{53}	2.22×10^{53}	3.46×10^{53}	5.20×10^{53}
^{59}Co	2.35×10^{53}	1.57×10^{53}	3.22×10^{53}	7.44×10^{52}	2.60×10^{53}	1.53×10^{53}	2.29×10^{53}
^{58}Ni	6.16×10^{53}	5.56×10^{53}	2.69×10^{55}	5.91×10^{53}	6.52×10^{53}	8.55×10^{53}	9.91×10^{53}
^{60}Ni	2.16×10^{54}	2.00×10^{54}	1.64×10^{54}	1.55×10^{54}	2.16×10^{54}	3.15×10^{53}	4.55×10^{53}
^{61}Ni	5.09×10^{52}	5.98×10^{52}	3.19×10^{53}	6.79×10^{52}	7.29×10^{52}	6.65×10^{52}	1.17×10^{53}
^{62}Ni	4.25×10^{52}	2.87×10^{52}	3.69×10^{54}	1.09×10^{53}	1.31×10^{53}	2.10×10^{53}	3.61×10^{53}
^{64}Ni	1.31×10^{52}	4.34×10^{51}	8.00×10^{52}	3.34×10^{52}	1.18×10^{53}	2.34×10^{53}	5.09×10^{53}
^{63}Cu	1.26×10^{52}	6.43×10^{51}	4.34×10^{52}	1.62×10^{52}	5.11×10^{52}	6.26×10^{52}	1.25×10^{53}
^{65}Cu	4.56×10^{51}	1.81×10^{51}	2.54×10^{52}	1.22×10^{52}	3.80×10^{52}	8.31×10^{52}	1.74×10^{53}
^{64}Zn	1.37×10^{53}	1.39×10^{53}	2.77×10^{52}	7.23×10^{52}	1.55×10^{53}	3.57×10^{52}	6.95×10^{52}
^{66}Zn	8.05×10^{51}	5.58×10^{51}	1.03×10^{53}	2.05×10^{52}	5.49×10^{52}	1.27×10^{53}	2.61×10^{53}
^{67}Zn	9.45×10^{50}	3.84×10^{50}	6.92×10^{51}	2.69×10^{51}	1.03×10^{52}	2.18×10^{52}	4.92×10^{52}
^{68}Zn	4.22×10^{51}	1.82×10^{51}	3.43×10^{52}	1.40×10^{52}	5.25×10^{52}	1.44×10^{53}	3.16×10^{53}
^{70}Zn	1.34×10^{50}	1.07×10^{50}	1.27×10^{51}	1.11×10^{50}	1.99×10^{50}	5.49×10^{50}	6.97×10^{50}
^{69}Ga	6.06×10^{50}	2.08×10^{50}	3.45×10^{51}	1.86×10^{51}	6.70×10^{51}	1.92×10^{52}	4.04×10^{52}
^{71}Ga	3.65×10^{50}	2.46×10^{50}	3.46×10^{51}	1.40×10^{51}	5.65×10^{51}	1.59×10^{52}	3.50×10^{52}
$^{140}\chi$	4.08×10^{57}	5.26×10^{57}	7.12×10^{57}	6.61×10^{57}	8.67×10^{57}	1.20×10^{58}	2.05×10^{58}
$^{100}\rho$	6.23×10^{55}	7.19×10^{55}	8.62×10^{55}	9.58×10^{55}	1.20×10^{56}	1.44×10^{56}	1.92×10^{56}

Table 11 Number of nucleons of each isotope of each star when the Metallicity equals 0.02

$M_*(M_\odot)$	13	15	18	20	25	30	40
^1H	7.37×10^{57}	8.13×10^{57}	9.02×10^{57}	9.50×10^{57}	1.01×10^{58}	1.05×10^{58}	4.25×10^{57}
^2H	9.99×10^{42}	1.22×10^{43}	4.98×10^{41}	1.44×10^{42}	1.18×10^{42}	1.29×10^{42}	9.23×10^{40}
^3He	2.35×10^{53}	2.61×10^{53}	2.75×10^{53}	2.85×10^{53}	2.65×10^{53}	2.54×10^{53}	6.05×10^{52}
^4He	5.15×10^{57}	6.29×10^{57}	7.32×10^{57}	8.10×10^{57}	8.67×10^{57}	1.00×10^{58}	5.64×10^{57}
He_{nuc}	1.23×10^{57}	1.76×10^{57}	1.89×10^{57}	2.06×10^{57}	1.13×10^{57}	9.58×10^{56}	-6.4×10^{57}
^6Li	1.58×10^{40}	3.80×10^{40}	3.33×10^{38}	1.31×10^{35}	1.14×10^{35}	7.07×10^{34}	4.16×10^{35}
^7Li	6.74×10^{46}	8.49×10^{44}	3.35×10^{44}	5.17×10^{45}	8.30×10^{44}	5.19×10^{44}	6.85×10^{44}
^9Be	5.52×10^{38}	1.86×10^{39}	9.06×10^{39}	5.65×10^{34}	2.67×10^{37}	5.89×10^{34}	8.42×10^{34}
^{10}B	1.34×10^{47}	1.63×10^{47}	1.70×10^{47}	1.76×10^{47}	1.81×10^{47}	1.84×10^{47}	8.34×10^{43}
^{11}B	5.13×10^{47}	7.21×10^{47}	7.68×10^{47}	7.83×10^{47}	8.11×10^{47}	8.29×10^{47}	3.86×10^{43}
^{12}C	1.28×10^{56}	7.80×10^{55}	1.63×10^{56}	2.93×10^{56}	1.82×10^{56}	2.99×10^{56}	7.14×10^{56}
^{13}C	1.20×10^{54}	1.38×10^{54}	1.65×10^{54}	1.74×10^{54}	7.98×10^{55}	2.30×10^{54}	5.16×10^{53}
^{14}N	5.75×10^{55}	7.37×10^{55}	7.92×10^{55}	8.61×10^{55}	1.01×10^{56}	1.22×10^{56}	6.96×10^{55}
^{15}N	4.90×10^{52}	7.15×10^{52}	1.83×10^{52}	2.68×10^{53}	5.55×10^{55}	7.85×10^{51}	6.85×10^{51}
^{16}O	2.61×10^{56}	1.94×10^{56}	9.22×10^{56}	1.26×10^{57}	2.81×10^{57}	3.86×10^{57}	8.78×10^{57}
^{17}O	1.04×10^{54}	9.73×10^{53}	1.05×10^{54}	1.14×10^{54}	1.61×10^{54}	2.02×10^{54}	1.16×10^{54}
^{18}O	4.19×10^{54}	3.04×10^{54}	1.40×10^{55}	6.25×10^{54}	1.01×10^{56}	7.40×10^{54}	1.47×10^{55}
^{19}F	1.70×10^{52}	1.96×10^{52}	5.35×10^{51}	7.25×10^{52}	1.43×10^{53}	9.35×10^{51}	6.30×10^{51}
^{20}Ne	4.16×10^{55}	4.06×10^{55}	1.78×10^{56}	4.72×10^{56}	1.02×10^{57}	1.12×10^{57}	2.65×10^{57}
^{21}Ne	2.69×10^{53}	1.56×10^{53}	2.42×10^{53}	2.17×10^{54}	1.90×10^{54}	3.56×10^{54}	5.95×10^{54}
^{22}Ne	5.41×10^{54}	2.06×10^{54}	7.94×10^{54}	1.08×10^{55}	2.01×10^{55}	2.23×10^{55}	1.31×10^{55}
^{23}Na	1.11×10^{54}	1.27×10^{54}	3.51×10^{54}	2.01×10^{55}	2.23×10^{55}	4.16×10^{55}	9.29×10^{55}
^{24}Mg	3.02×10^{55}	4.54×10^{55}	1.23×10^{56}	8.58×10^{55}	2.61×10^{56}	2.25×10^{56}	3.71×10^{56}
^{25}Mg	3.07×10^{54}	1.75×10^{54}	8.48×10^{54}	1.72×10^{55}	3.75×10^{55}	3.74×10^{55}	8.72×10^{55}
^{26}Mg	2.59×10^{54}	2.06×10^{54}	7.02×10^{54}	1.06×10^{55}	3.26×10^{55}	3.34×10^{55}	8.78×10^{55}
^{26}Al	2.55×10^{52}	9.14×10^{51}	4.42×10^{52}	1.80×10^{52}	1.04×10^{53}	4.67×10^{52}	7.95×10^{52}
^{27}Al	1.80×10^{54}	2.92×10^{54}	1.20×10^{55}	1.19×10^{55}	3.23×10^{55}	4.08×10^{55}	9.94×10^{55}
^{28}Si	8.96×10^{55}	1.00×10^{56}	1.21×10^{56}	7.57×10^{55}	1.53×10^{56}	2.87×10^{56}	2.89×10^{56}
^{29}Si	1.78×10^{54}	2.64×10^{54}	8.34×10^{54}	2.44×10^{54}	8.46×10^{54}	8.84×10^{54}	1.20×10^{55}
^{30}Si	1.86×10^{54}	3.29×10^{54}	8.17×10^{54}	2.92×10^{54}	7.41×10^{54}	1.27×10^{55}	1.17×10^{55}
^{31}P	4.49×10^{53}	8.46×10^{53}	2.08×10^{54}	7.88×10^{53}	1.80×10^{54}	3.07×10^{54}	4.22×10^{54}
^{32}S	4.47×10^{55}	4.16×10^{55}	4.42×10^{55}	3.37×10^{55}	5.98×10^{55}	1.29×10^{56}	1.31×10^{56}
^{33}S	2.40×10^{53}	2.91×10^{53}	4.20×10^{53}	2.71×10^{53}	3.89×10^{53}	5.76×10^{53}	5.76×10^{53}
^{34}S	1.96×10^{54}	1.83×10^{54}	2.62×10^{54}	2.04×10^{54}	2.71×10^{54}	4.50×10^{54}	4.10×10^{54}
^{36}S	6.43×10^{51}	3.04×10^{51}	1.44×10^{52}	1.10×10^{52}	2.90×10^{52}	6.10×10^{52}	1.13×10^{53}
^{35}Cl	1.38×10^{53}	1.64×10^{53}	2.10×10^{53}	1.50×10^{53}	1.94×10^{53}	2.79×10^{53}	3.31×10^{53}
^{37}Cl	3.63×10^{52}	2.98×10^{52}	6.77×10^{52}	1.01×10^{53}	2.04×10^{53}	3.15×10^{53}	6.76×10^{53}
^{36}Ar	7.53×10^{54}	5.87×10^{54}	6.56×10^{54}	5.58×10^{54}	9.49×10^{54}	2.17×10^{55}	2.17×10^{55}
^{38}Ar	8.22×10^{53}	7.80×10^{53}	9.38×10^{53}	8.07×10^{53}	1.08×10^{54}	1.96×10^{54}	1.98×10^{54}
^{40}Ar	1.07×10^{51}	1.26×10^{51}	1.56×10^{51}	1.69×10^{51}	3.21×10^{51}	8.88×10^{51}	6.79×10^{51}
^{39}K	6.01×10^{52}	8.38×10^{52}	9.19×10^{52}	8.85×10^{52}	1.06×10^{53}	1.55×10^{53}	1.37×10^{53}
^{40}K	1.53×10^{50}	7.70×10^{49}	1.65×10^{50}	1.11×10^{50}	2.40×10^{50}	3.52×10^{50}	4.85×10^{50}

(continued)

Table 11 (continued)

$M_*(M_\odot)$	13	15	18	20	25	30	40
^{41}K	5.61×10^{51}	7.01×10^{51}	9.27×10^{51}	1.08×10^{52}	1.66×10^{52}	2.36×10^{52}	5.41×10^{52}
^{40}Ca	5.89×10^{54}	4.80×10^{54}	5.44×10^{54}	4.48×10^{54}	7.85×10^{54}	1.88×10^{55}	1.87×10^{55}
^{42}Ca	1.66×10^{52}	2.10×10^{52}	2.48×10^{52}	2.08×10^{52}	2.96×10^{52}	4.97×10^{52}	5.37×10^{52}
^{43}Ca	1.63×10^{51}	1.80×10^{51}	2.54×10^{51}	3.02×10^{51}	4.10×10^{51}	3.56×10^{51}	5.91×10^{51}
^{44}Ca	4.20×10^{52}	4.25×10^{52}	6.53×10^{52}	6.94×10^{52}	6.12×10^{52}	4.84×10^{52}	5.59×10^{52}
^{46}Ca	2.54×10^{50}	3.99×10^{50}	9.38×10^{50}	7.37×10^{50}	1.41×10^{51}	1.41×10^{51}	3.43×10^{51}
^{48}Ca	1.89×10^{51}	2.23×10^{51}	3.27×10^{51}	2.81×10^{51}	3.27×10^{51}	1.70×10^{52}	2.91×10^{51}
^{45}Sc	8.73×10^{50}	1.11×10^{51}	1.51×10^{51}	1.56×10^{51}	3.35×10^{51}	3.08×10^{51}	7.37×10^{51}
^{46}Ti	7.20×10^{51}	1.03×10^{52}	1.14×10^{52}	1.02×10^{52}	1.35×10^{52}	2.23×10^{52}	2.53×10^{52}
^{47}Ti	5.28×10^{51}	4.12×10^{51}	6.40×10^{51}	7.85×10^{51}	6.70×10^{51}	1.06×10^{52}	8.08×10^{51}
^{48}Ti	9.74×10^{52}	1.28×10^{53}	1.58×10^{53}	1.38×10^{53}	1.83×10^{53}	3.03×10^{53}	2.92×10^{53}
^{49}Ti	5.68×10^{51}	6.07×10^{51}	7.32×10^{51}	7.63×10^{51}	1.32×10^{52}	2.30×10^{52}	2.90×10^{52}
^{50}Ti	2.79×10^{51}	2.89×10^{51}	4.68×10^{51}	6.97×10^{51}	1.47×10^{52}	1.19×10^{52}	4.56×10^{52}
^{50}V	2.43×10^{49}	3.19×10^{49}	7.91×10^{49}	4.66×10^{49}	1.15×10^{50}	2.98×10^{50}	2.24×10^{50}
^{51}V	1.41×10^{52}	1.19×10^{52}	1.37×10^{52}	1.59×10^{52}	1.72×10^{52}	2.99×10^{52}	2.68×10^{52}
^{50}Cr	2.67×10^{52}	4.73×10^{52}	4.41×10^{52}	3.71×10^{52}	6.25×10^{52}	9.23×10^{52}	1.10×10^{53}
^{52}Cr	7.76×10^{53}	1.52×10^{54}	1.53×10^{54}	1.28×10^{54}	2.44×10^{54}	4.50×10^{54}	4.53×10^{54}
^{53}Cr	7.47×10^{52}	1.11×10^{53}	1.15×10^{53}	9.92×10^{52}	1.69×10^{53}	3.28×10^{53}	2.98×10^{53}
^{54}Cr	7.85×10^{51}	8.14×10^{51}	1.82×10^{52}	1.99×10^{52}	3.32×10^{52}	6.13×10^{52}	6.78×10^{52}
^{55}Mn	3.37×10^{53}	4.55×10^{53}	4.85×10^{53}	4.31×10^{53}	6.59×10^{53}	1.25×10^{54}	9.92×10^{53}
^{54}Fe	2.30×10^{54}	2.95×10^{54}	2.98×10^{54}	2.66×10^{54}	4.60×10^{54}	8.23×10^{54}	7.69×10^{54}
^{56}Fe	9.97×10^{55}	1.02×10^{56}	1.04×10^{56}	1.06×10^{56}	1.08×10^{56}	1.10×10^{56}	9.68×10^{55}
^{57}Fe	2.66×10^{54}	2.38×10^{54}	3.20×10^{54}	2.80×10^{54}	2.24×10^{54}	3.38×10^{54}	1.17×10^{54}
^{58}Fe	1.45×10^{53}	1.32×10^{53}	7.02×10^{53}	5.94×10^{53}	1.00×10^{54}	2.86×10^{54}	1.77×10^{54}
^{59}Co	1.69×10^{53}	1.08×10^{53}	2.18×10^{53}	3.50×10^{53}	4.66×10^{53}	6.46×10^{53}	8.31×10^{53}
^{58}Ni	2.67×10^{54}	1.37×10^{54}	3.23×10^{54}	2.20×10^{54}	1.87×10^{54}	1.86×10^{54}	1.06×10^{54}
^{60}Ni	2.55×10^{54}	2.24×10^{54}	2.54×10^{54}	2.98×10^{54}	2.17×10^{54}	7.33×10^{53}	1.44×10^{54}
^{61}Ni	9.83×10^{52}	1.49×10^{53}	1.25×10^{53}	1.78×10^{53}	2.59×10^{53}	1.32×10^{53}	4.68×10^{53}
^{62}Ni	2.72×10^{53}	1.96×10^{53}	5.56×10^{53}	4.62×10^{53}	7.92×10^{53}	2.66×10^{53}	1.40×10^{54}
^{64}Ni	4.01×10^{52}	2.06×10^{52}	3.58×10^{52}	2.24×10^{53}	7.38×10^{53}	8.52×10^{52}	2.66×10^{54}
^{63}Cu	2.19×10^{52}	1.56×10^{52}	2.28×10^{52}	1.53×10^{53}	3.52×10^{53}	1.03×10^{53}	8.20×10^{53}
^{65}Cu	1.23×10^{52}	8.60×10^{51}	9.13×10^{51}	5.27×10^{52}	1.75×10^{53}	1.56×10^{52}	7.77×10^{53}
^{64}Zn	1.16×10^{53}	8.11×10^{52}	6.88×10^{52}	1.44×10^{53}	6.43×10^{52}	2.24×10^{52}	2.01×10^{53}
^{66}Zn	2.08×10^{52}	2.13×10^{52}	2.60×10^{52}	7.50×10^{52}	2.86×10^{53}	1.84×10^{52}	1.28×10^{54}
^{67}Zn	3.45×10^{51}	2.13×10^{51}	2.43×10^{51}	1.55×10^{52}	6.30×10^{52}	3.26×10^{51}	2.97×10^{53}
^{68}Zn	1.58×10^{52}	8.97×10^{51}	1.40×10^{52}	7.22×10^{52}	3.76×10^{53}	1.52×10^{52}	1.86×10^{54}
^{70}Zn	2.19×10^{51}	6.70×10^{50}	6.14×10^{50}	1.94×10^{51}	1.01×10^{52}	4.42×10^{50}	3.17×10^{52}
^{69}Ga	2.00×10^{51}	1.01×10^{51}	1.09×10^{51}	7.98×10^{51}	3.28×10^{52}	1.65×10^{51}	2.10×10^{53}
^{71}Ga	1.77×10^{51}	9.53×10^{50}	1.27×10^{51}	6.16×10^{51}	3.34×10^{52}	1.32×10^{51}	1.58×10^{53}
$^{140}\chi$	4.25×10^{57}	5.55×10^{57}	6.92×10^{57}	8.01×10^{57}	1.16×10^{58}	1.55×10^{58}	3.38×10^{58}
$^{100}\rho$	3.11×10^{56}	3.59×10^{56}	4.31×10^{56}	4.79×10^{56}	5.99×10^{56}	7.19×10^{56}	9.58×10^{56}

References

Amado A, Batista C, Campos PR (2018) A theoretical approach to the size-complexity rule. Evolution 72(1):18–29

Aristotle's Generation of Animals (2018) In: Falcon A, Lefebvre D (eds) Aristotle's generation of animals: a critical guide. Cambridge critical guides. Cambridge University Press, Cambridge, p. I

Avellar MGBd, Souza RAd, Horvath JE (2016) Trends of stellar entropy along stellar evolution. Res Astron Astrophy 16(2):021

Bejan A, Lorente S, Yilbas B, Sahin A (2011) The effect of size on efficiency: Power plants and vascular designs. Int J HeatMass Transf 547–8:1475–1481

Bell G, Mooers AO (1997) Size and complexity among multicellular organisms. Biol J Linnean Soc 60(3):345–363

Bonner JT (1995) The evolution of life's complexity. Nature 374(6522):508–509

Bonner JT (2004) Perspective: the size-complexity rule. Evolution 58(9):1883–1890

Bonner JT (2015) Size and cycle: an essay on the structure of biology, vol 2087. Princeton University Press, Princeton

Carneiro RL (1967) On the relationship between size of population and complexity of social organization. Southwestern J Anthropol 23(3):234–243

Chaisson, EJ (2002) Cosmic evolution. Harvard University Press, Cambridge

Curtis S, Ebinger K, Fröhlich C, Hempel M, Perego A, Liebendörfer M, et al. (2018) Pushing core-collapse supernovae to explosions in spherical symmetry. iii. nucleosynthesis yields. Astrophys J 870(1):2

Einstein A (1905) On the electrodynamics of moving bodies [Annalen der Physik 17 (1905)]. The Collected Papers of Albert Einstein, vol2. Createspace, Scotts Valley

Frebel A, Aoki W, Christlieb N, Ando H, Asplund M, Barklem PS, et al. (2005). Nucleosynthetic signatures of the first stars. Nature 434(7035):871

Frebel A, Norris JE, Aoki W, Honda S, Bessell MS, Takada-Hidai M, et al. (2007). Chemical abundance analysis of the extremely metal-poor star he 1300+ 0157. Astrophys J 658(1):534

Georgiev GY, Chatterjee A (2016) The road to a measurable quantitative understanding of self-organization and evolution. In: Evolution and transitions in complexity. Springer, Berlin, pp 223–230

Georgiev G, Georgiev I (2002) The least action and the metric of an organized system. Open Syst Inf Dyn 9(4):371

Georgiev GY, Henry K, Bates T, Gombos E, Casey A, Daly M, et al (2015) Mechanism of organization increase in complex systems. Complexity 21(2):18–28

Georgiev GY, Gombos E, Bates T, Henry K, Casey A, Daly M (2016) Free energy rate density and self-organization in complex systems. In: Proceedings of ECCS 2014. Springer, Berlin, pp 321–327

Georgiev GY, Chatterjee A, Iannacchione G (2017) Exponential self-organization and Moore's law: measures and mechanisms. Complexity 2017:8170632

Hegel GWF (2014) Georg Wilhelm Friedrich Hegel: the science of logic. Cambridge University Press, Cambridge

Johnson CI, Pilachowski CA (2010) Chemical abundances for 855 giants in the globular cluster Omega Centauri NGC 5139). Astrophy J 722(2):1373–1410. https://doi.org/10.1088%2F0004-637x%2F722%2F2%2F1373

Kleiber M, et al (1932) Body size and metabolism. Hilgardia 611:315–353

Maurer BA (1996) Species diversity in space and time. Ecology 77(4):1314–1316

McCarthy MC, Enquist BJ (2005) Organismal size, metabolism and the evolution of complexity in metazoans. Evol Ecol Res 7(5):681–696

Mori K, Famiano MA, Kajino T, Suzuki T, Garnavich PM, Mathews GJ, et al (2018) Nucleosynthesis constraints on the explosion mechanism for type ia supernovae. Astrophys J 863(2):176

Nomoto K (2016) First stars, hypernovae, and superluminous supernovae. Int J Modern Phys D 25(10):1630025

Nomoto K, Suzuki T (2013) Supernova yields for chemical evolution modeling. Proce Int Astronom Union 9(S298):154–166

Nomoto K, Hashimoto M, Tsujimoto T, Thielemann F-K, Kishimoto N, Kubo Y, et al (1997) Nucleosynthesis in type ii supernovae. Nuclear Phys A 616(1–2):79–90

Nomoto K, Tominaga N, Umeda H, Kobayashi C (2005) Yields of population iii supernovae and the abundance patterns of extremely metal-poor stars. Proce Int Astronom Union 1(S228):287–296

Nomoto K, Tominaga N, Umeda H, Kobayashi C, Maeda K (2006) Nucleosynthesis yields of core-collapse supernovae and hypernovae, and galactic chemical evolution. Nuclear Phys A 777:424–458

Nomoto K, Kobayashi C, Tominaga N (2013) Nucleosynthesis in stars and the chemical enrichment of galaxies. Ann Rev Astron Astrophys 51:457–509

Pignatari M, Herwig F, Hirschi R, Bennett M, Rockefeller G, Fryer C, et al (2016) Nugrid stellar data set. i. stellar yields from h to bi for stars with metallicities z= 0.02 and z= 0.01. Astrophys J Suppl Ser 225(2):24

Prantzos N, Abia C, Limongi M, Chieffi A, Cristallo S (2018) Chemical evolution with rotating massive star yields–i. the solar neighbourhood and the s-process elements. Monthly Notices Roy Astronom Soc 476(3):3432–3459

Ritter C, Herwig F, Jones S, Pignatari M, Fryer C, Hirschi R (2018) Nugrid stellar data set–ii. stellar yields from h to bi for stellar models with m zams= 1–25 m and z= 0.0001–0.02. Monthly Notices Roy Astronom Soc 480(1):538–571

Sinha S, Fröhlich C, Ebinger K, Perego A, Hempel M, Eichler M, et al (2017) Pushing core-collapse supernovae to explosions in spherical symmetry: nucleosynthesis yields. In: Proceedings of the 14th international symposium on nuclei in the cosmos (nic2016), p 020608

Steigman G (2007) Primordial nucleosynthesis in the precision cosmology era. Annu Rev Nucl Part Sci 57:463–491

Suda T, Katsuta Y, Yamada S, Suwa T, Ishizuka C, Komiya Y, et al (2008) Stellar abundances for the galactic archeology (saga) database—compilation of the characteristics of known extremely metal-poor stars. Public Astronom Soc Jpn 60(5):1159–1171

Takahashi K, Umeda H, Yoshida T (2014) Stellar yields of rotating first stars. i. yields of weak supernovae and abundances of carbon-enhanced hyper-metal-poor stars. Astrophys J794(1):40

Thielemann F-K, Diehl R, Heger A, Hirschi R, Liebendörfer M (2018) Massive stars and their supernovae. In: Astrophysics with radioactive isotopes. Springer, Berlin, , pp 173–286

Tominaga N, Iwamoto N, Nomoto K (2014) Abundance profiling of extremely metal-poor stars and supernova properties in the early universe. Astrophys J 785(2):98

Umeda H, Nomoto K (2003) First-generation black-hole-forming supernovae and the metal abundance pattern of a very iron-poor star. Nature 422(6934):871.

Wagoner RV, Fowler WA, Hoyle F (1967) On the synthesis of elements at very high temperatures. Astrophys J 148:3

Walker SI (2019) The new physics needed to probe the origins of life. Nature 569(7754):36–39

Wanajo S, Nomoto K, Janka H-T, Kitaura F, Müller B (2009) Nucleosynthesis in electron capture supernovae of asymptotic giant branch stars. Astrophys J 695(1):208

West GB (2017) Scale: the universal laws of growth, innovation, sustainability, and the pace of life in organisms, cities, economies, and companies. Penguin, London

West GB, Brown JH, Enquist BJ (1999) The fourth dimension of life: fractal geometry and allometric scaling of organisms. Science 2845420:1677–1679

Trends and Technical Advancements on High-Efficiency Electric Motors: A Review

Jawad Faiz and Farbod Parvin

Abstract In order to reduce the global energy consumption, the energy efficiency must be improved. Electric motors are the major consumers of electrical energy and their efficiency improvement can have a very large impact on saving electrical energy. This chapter focuses on the latest advancements of various types of electric motors. These advancements spring from the high-end materials, new structures, or improved construction techniques. The chapter classifies electric motors by their types in different sections and the latest trends and advancements of each specified motor are discussed thoroughly. Finally, a brief comparison is conducted using the related literature and future possibilities of different types of motors.

Keywords Efficiency · Induction machines · Losses reduction · Optimal utilization · Electrical motors

1 Introduction

Energy awareness is considered to be one of the most important motives in engineering researches, and electric motors are no exception. In fact, improving electric motors' efficiency is a top priority in this field. Motor-related systems consume over 60% of electricity worldwide and they are the largest consumers of electric power (Lu 2016). Therefore, efficiency improvement of such systems, even below 1%, can make a drastic saving in the electrical energy demand. Because of the multidisciplinary nature of the electric motor field, efficiency improvement can be realized by many ways. Engineering solutions can be electrical, mechanical, material, or even physics based (as in high-temperature superconductor) and all can be applicable here.

J. Faiz (✉) · F. Parvin
Center of Excellence on Applied Electromagnetic Systems, School of Electrical and Computer Engineering, College of Engineering, University of Tehran, Tehran, Iran
e-mail: jfaiz@ut.ac.ir

G. Y. Georgiev, M. Shokrollahi-Far (eds.), *Efficiency in Complex Systems*, Springer Proceedings in Complexity, https://doi.org/10.1007/978-3-030-69288-9_5

From electrical engineering point of view, efficiency improvement depends on the motor type. This rather diverse category consists of some well-known motors such as induction motor (IM), direct current motor (DCM), permanent magnet synchronous motor (PMSM), and switched reluctance motor (SRM) to more modern and advanced structures such as brushless DC motor (BLDC), synchronous reluctance motor (SynR), segmented SRM (SSRM), and flux switching motor (FSM). Research is still ongoing for nearly all of the abovementioned motors (except DCM) but some of the modern structures are still far from a mature technology.

The aim of this chapter is not only to identify the latest methods of efficiency improvement, but also to classify them and trace them back to their origins. As it will be explained later, different techniques can benefit the machine performance by many ways. Considering copper loss for instance, it can be lowered by using cast-copper rotor in IMs (Lu 2016) or it can be degraded by utilizing fractional-slot concentrated windings in synchronous motor. This review approach has the benefit of identifying all of the latest trends in the field and tracing them back to their origins and see if it is possible to derive another method or technique based on their own origins. A brief summary of reviewed articles has been given in Table 1.

Due to the underlying differences in different kind of motors, various types and advances in each structure have been reported and discussed thoroughly in this chapter.

2 Conventional and Segmented Switched Reluctance Motors

Unlike SynR motors, rotor geometry of conventional SRMs has not been developed significantly over the past 30 years (Mecrow et al. 2002). Nowadays, the SSRM is considered as an improved version of SRM and its advantages over its predecessor will be discussed here. Aside from its manufacturing difficulty (Xu et al. 2016), the SSRM performance in every aspect is superior to the conventional SRM. A tooth-wound SSRM has been proposed by Widmer et al. (2015) as an alternative for traction application and has been optimized as a close competitor of the interior permanent magnet (IPM) motor of the Nissan Leaf car. Since the power-to-volume ratio of the SSRMs is lower than that of the IPM motors, no size constraint has been considered. However, the traction application requires the lowest possible mass. As a result of such requirement, one of the major aspects of this electric motor is its tooth-wound winding, in spite of the other SSRMs. Obviously, another advantage of such winding is lower copper losses and higher efficiency. Strengths and weaknesses of the SSRM have been investigated more thoroughly by Mecrow et al. (2004). Two types of the SSRM with two winding layouts of full pitched and single tooth have been compared with the conventional SRM and a BLCD motor. It has been reported that the whole-coiled motor has a better magnetic utilization; that is, more percentage of the stator core participates in the magnetic circuit of each phase. The other design which has a similar winding with the conventional SRM is lighter since

Table 1 Recent advancements and research trends in various types of motors

SRM			PM motor	
SRM	SSRM	IM	PMSM	FSPM
• Thin ribs for noise reduction	• Displaced segments • Aluminum rotor • Circular slots • Single-tooth winding	• Copper rotor • Low-harmonic windings • Slitted solid rotor • Core lengthening • Rotor replacement • 1-phase utilization of 3-phase motors	• Amorphous alloy core • Phase group concentrated • Sinusoid PM shape • Current harmonic injection • FSCW • Asymmetrical rotor structure	• Flux diverters • Enhanced demagnetization • Hybrid excitation

it has less copper but requires certain design considerations such as stator tooth width.

A novel SSRM (Oyama et al. 2006) has an aluminum rotor, which embodies the segments of the rotor. Such structure can fulfill two requirements. First is the mechanical robustness and ease of manufacturing. Second, which is of more interest here, is the improved torque of the machine, because it has an additional eddy current component. Lower radial force is another advantage of the SSRM, because the air gap flux path is more circumferential rather than radial. A higher number of rotor segments has been suggested in the work of Vandana and Fernandes (2015) for high efficiency and low copper loss, especially in direct-drive applications. However, increasing the number of segments will also increase the core losses of the machine; however, it does not recommend structures with more than 16 segments, because efficiency improvement is insignificant.

One of the intrinsic advantages of the SSRM over the SRM has been introduced by Vattikuti et al. (2008). A better magnetic utilization allows more compact structures. By optimizing this feature in the SSRM, a circular-slot SSRM has been recommended. Such geometry confines the flux to circular paths and removes the need of conventional stator back-iron. The result is a more compact and efficient motor in exchange of higher resistant windings.

Torque ripple can be reduced in a dual-axial SSRM (Madhavan and Fernandes 2014) with displaced rotor segment technique. By increasing the $\frac{dL}{d\theta}$ in the incoming phase, this technique makes the commutation transition more smooth and removes the dips in the torque profile of the motor. In contrast to other SSRMs, a stator-segmented SRM with an outer rotor has been proposed by Mousavi-Aghdam et al. (2017). The geometry of this machine allows the designer to use concentrated winding in the stator and reduce the weight of the active material. Short magnetic path and flux reverse free of its stator segments make an ideal solution for its core loss reduction.

In parallel with the SSRM, development is still ongoing for the conventional SRM structures. A major challenge of SRMs has been addressed by Kiyota et al. (2016). The acoustic noise, windage loss, and vibration associated with the salient structure of the SRM rotor have been significantly reduced by a series of ultrathin ribs connecting the adjacent poles of the rotor. Design considerations of this topology are unaligned inductance and rotor mechanical robustness. Eventually, an estimated efficiency improvement has been reported.

3 Induction Motors

Economically, efficiency improvement of electric motors is the most important factor in the performance evaluation. Reliability is also another important aspect which can be taken into account in the efficient motors. In two motors with the same insulation material class, the one with higher efficiency has lower temperature rise, which means its insulation material aging is slower. Moreover, bearings and lubricants perform better and longer when less heat is generated (de Almeida et al. 2009).

A no tooling cost solution has been introduced by Alberti et al. (2014) for efficiency improvement of IMs. It suggests that simply lengthening the active part of the motor can improve the efficiency of the motors. This argument can be stated reversely; that is, utilizing a larger motor for a lower power means that the magnetic and electric loading of the electric machine is lower than its threshold, and consequently, the machine losses will reduce. Some suggested remanufacturing of industrial IMs (Ni et al. 2016; Li et al. 2017). In the work of Li et al. (2017), synchronous reluctance rotor replacement has been reported. In the work of Ni et al. (2016), rotor of an IM has been replaced with rotor of IPM motor. By utilizing maximum efficiency per ampere control algorithm, the efficiency class has been increased from IE2 to IE4. In the work of Jang (2017), aluminum segments have been placed in the barriers of a SynR motor in order to improve the efficiency. This structure has its own downsides such as starting performance and out-of-step instability.

Most studies on IMs have been concentrated around the cast-copper rotors (Finley and Hodowanec 2001; Goss et al. 2013; Malinowski et al. 2004; Dorrell 2014; Rajkumar et al. 2017; Lin and Hwang 2016). It has been emphasized (Finley and Hodowanec 2001) that simply using a cast-copper rotor may not be satisfactory, because the rotor bar resistance is low causing the low blocked-rotor torque. A copper rotor IM has been designed for traction application (Goss et al. 2013), which is comparable with the Toyota Prius IPM motor. Generally IM has lower efficiency (even with copper-made rotor), a bigger and heavier motor, higher volt-ampere rating of the inverter, and bigger batteries that are inevitable outcomes of such a design. Therefore, such a structure is only recommended for hybrid drivetrains or in the cases where initial cost matters. The cast-copper rotors can improve the overall efficiency of IM up to 1% or 2% (Malinowski et al. 2004). This number may differ depending on the size of the motor (lower for larger motors).

The stray load loss is notably lower in low-slip operation (Dorrell 2014). So, as in Fig. 1, lower rotor resistance is desirable in terms of stray load loss; however, this affects the starting performance of the motor. It has also been reported that removal of the fins at both ends of the rotor drastically reduces the windage loss. A comparative analysis has been conducted by Rajkumar et al. (2017) regarding the material selection of the rotor bars and the end rings. The results suggest that aluminum bars and copper-end rings make the most torque-dense structures. On the other hand, IMs with copper bars (either aluminum or copper rings) have superior efficiency. A more reliable six-phase copper rotor IM has been optimized by Lin and Hwang (2016). A multiphase structure has many benefits such as improved reliability, lower torque ripple, and higher efficiency. Bottleneck of such motors is the manufacturing cost, so a compromised optimization between the mentioned advantages and manufacturing cost can be made.

One of the interesting trends in IM efficiency improvement is single-phase utilization of three-phase motors. By interchanging the winding connections and using capacitors (as phase shifters) between these connections, a three-phase

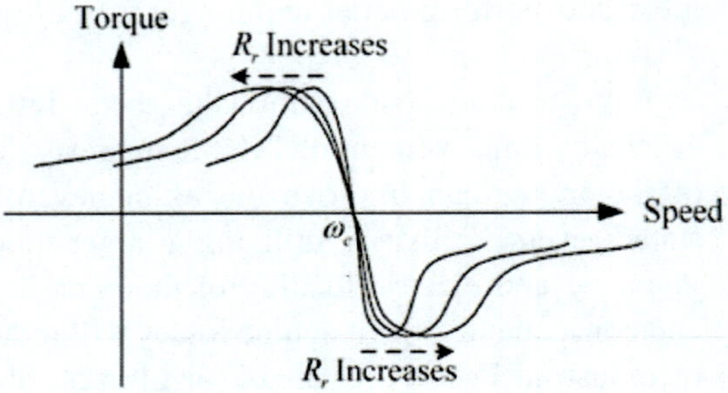

Fig. 1 Impact of increasing rotor resistance on the slip of induction motor

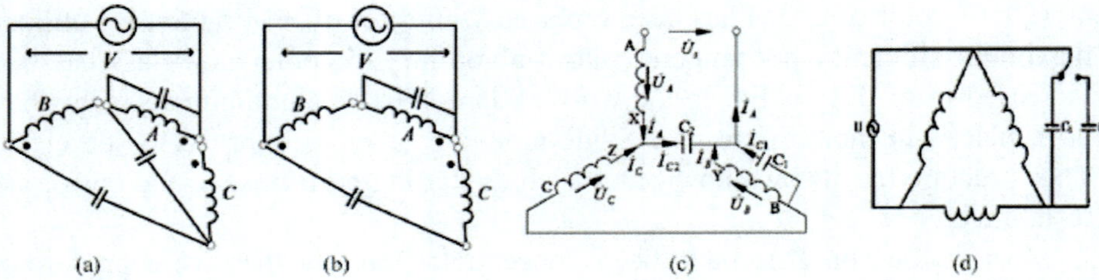

Fig. 2 Single-phase utilization of three-phase IM; (**a**) Smith connection; (**b**) SEMIHEXTH connection; (**c**) connection proposed by Wang et al. (2010); (**d**) connection proposed by Gonzalez et al. (2014)

machine can operate by a single-phase power supply (Gonzalez et al. 2014). Numerous techniques and topologies have been reported so far. Figure 2 presents a number of these configurations. There is another topology that has only two capacitors (Wang et al. 2010). As shown in Fig. 2c, three windings are in series and capacitors are connected in parallel with winding B and windings B and C, respectively. A parallel three-phase winding with two series-connected switchable capacitors can also be recommended (Gonzalez et al. 2014) (Fig. 2d). One pair of capacitors has been considered for the steady-state operation while the other one has been optimized for startup operation.

In addition to the abovementioned research trends, some studies have proposed solutions that are rather unconventional. A megawatt high-speed solid rotor IM has been designed by Zhang et al. (2017). Important feature of a solid rotor is that it enhances thermal conductivity, so the heat can be dissipated more effectively. It uses axially slitted rotor which has some advantages such as ease of penetration of the main component of the flux into the rotor and suppressing the eddy current on the rotor surface. In addition, two copper rings at two ends of the rotor provide high-current conduction paths. An interesting approach has been adopted by Zhang et al. (2014) based on the winding configuration which reduces the stray loss of the machine. Concentric low-harmonic winding and wye-delta mixed connection are the two proposed methods.

4 Permanent Magnet Synchronous Motors

Amorphous alloy (AA) stator core is the main focus in the work of Fan et al. (2014). Laminated steel can be substituted by AA. The AA must be used with care and taking into account its lower saturation level. With a proper optimization, a more compact V-shaped IPM motor has been developed compared to the original steel-laminated motor. Performance of an amorphous made SPM has been investigated by Tong et al. (2016). It has been shown that if the saturation has been taken into consideration, the efficiency can be even higher at high switching frequencies. Two prototypes have been reported by Jang (2017): one is an axial 11 kW motor developed by Hitachi that achieved 96% efficiency and the other one has been manufactured in the University of Adelaide using the water-jetting method.

Because of the intrinsic saliency of IPM motors, torque ripple reduction is a more challenging task. A set of design variables related to magnets and barriers in V-shaped IPM motors have been optimized in the work of Kim et al. (2009). As their objective functions are mostly the torque ripple, these ripples have been significantly reduced. A similar technique has been used for both increasing the torque of ferrite PM motors and reducing their torque ripples (Zhao et al. 2014a, 2015a, b, 2017a, b). Since the ferrite PMs are weak in terms of remnant flux density (B_r), their flux should be focused on air gap. The "concentrated phase-group" winding does so by attracting the flow of flux into one phase group. However, this method causes serious fluctuation in the torque profile. So, in order to minimize this cogging effect, dual-air gap structures have been suggested. Either with two stators or two rotors, they are displaced with respect to each other to alternately fill the void of no-torque areas.

Another technique has been proposed by Zhao et al. (2014b) and Zhao et al. (2015c) for low-ripple applications. In contrast to the common method of skewing, a "sinusoidal PM" shaping method has been suggested (Fig. 3). This method provides a more sinusoidal back EMF and does not have the axial force problem of the common skewing methods. It should be mentioned that since the field distribution is more sinusoidal, core loss will be reduced significantly. The ratio of pole-arc to pole-pitch, duct shape, and saliency are other parameters that have been optimized by Kim et al. (2007) for enhanced torque capability of the IPM motors. Compared to spoke-type IPMs, flared-shape IPMs (Yoon and Kwon 2016) can be a superior solution in terms of torque and demagnetization, but inferior in terms of efficiency. Since the interaction of the non-sinusoidal back EMF and sinusoidal phase current causes the torque ripple, Lee et al. (2008) have suggested that injecting a suitable set of current harmonics can improve the torque performance of the IPM motor. With a similar argument and in a similar motor, some harmonics can be injected into the "shape" of the rotor poles (Wang et al. 2014; Liu et al. 2018). While the former has done this by injecting third cosine harmonic into the rotor pole of an IPM motor, the latter has injected third and fifth harmonics into the shape of the magnets in a surfaced-mounted permanent magnet (SPM0 motor). As expected, both methods have led to a lower torque ripple.

Fractional-slot concentrated winding (FSCW) is defined as follows:

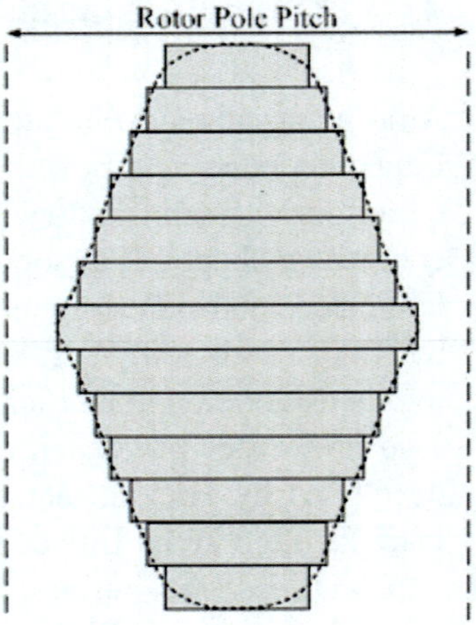

Fig. 3 PM sinusoidal shape and its sine-wave approximation

Table 2 Advantages and disadvantages of FSCW

Advantages	Disadvantages
• High slot fill factor	• High space harmonic
• Increased slot thermal conductivity	• Increased torque ripple
• Short end windings	• Increased iron loss
• Stator segmentation and ease of manufacturing	• Decreased power factor
	• Limitations on slot-pole combinations

$$q = \frac{Q_s}{6p} = \frac{z}{n} = \begin{cases} \text{Fractional} \leq 1, & \text{FSCW} \\ \text{Fractional} \ > 1, & \text{FSDW} \\ \text{Positive Integer, ISDW} \end{cases} \tag{1}$$

A comprehensive review on FSCW in SynR motors can be found in the work of Spargo et al. (2015). For the sake of brevity, positive and negative points of this winding layout have been listed in Table 2. A thorough performance analysis on FSCW motor has been conducted by Min et al. (2018). While the advantages of this winding layout such as low copper loss and improved reliability are well known, many of its performance characteristics should be calculated with FEM. Analytical expressions for the back EMF, inductance, and cogging torque have been given.

One of the most innovative ideas in PM motors has been introduced by Zhao et al. (2014c) and Zhao et al. (2017c). By using special asymmetry in rotor structures, reluctance torque and PM torque components of the proposed IPM motors reach their maximum at the same current phase angle. These topologies can be very attractive in the cases of ferrite PM or SynR motors where torque density is lower.

The aspect ratio and its effect on the efficiency of electric motors have been discussed by Tsunata et al. (2017). In motors with high aspect ratio, i.e., taller ones, conventional radial structures are satisfactory but in low-aspect-ratio structures with flatter shapes, the air gap surface is very low and end winding is a huge fraction

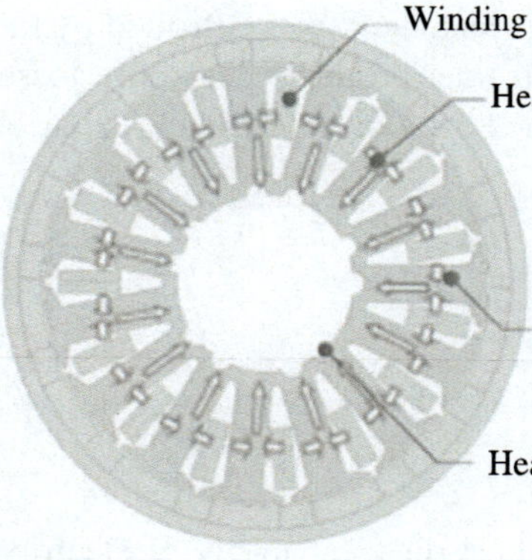

Fig. 4 Utilization of heat pipes according to heat flow path

of the overall length. As a result, torque density is low, and the axial flux motors are preferred. In addition, bonded magnet has been used which has high resistance leading to lower magnet eddy current loss and higher efficiency.

Thermal performance of the PM machines is an important problem (Li et al. 2016). It is clear that high efficiency and high power density are two contradicting demands. High-power motor means that either its voltage or its current should be high. Due to the limitation of bus voltage, often a higher current rating is chosen which imposes limitations on the electric motor efficiency. The heat pipes have been suggested as an efficient heat removal system. The heat exchange path and pipe location have been shown in Fig. 4.

In traction applications, especially electric vehicles (EVs), efficiency has broader meaning. There are certain conditions in this application such as direct-drive power train, operation in a wide range of loads and speeds, and short-duty capability that are not present in other applications. The following are concerned with the efficiency improvement of motors in these conditions.

Since most of the motors operate in speeds higher than their actual need of application, efficiency improvement in direct-drive applications is a challenging topic. The PM Vernier motor may be proposed for such applications (Xu et al. 2015). Comparison of FSCW and integral-slot distributed winding (ISDW) for PMV electric motors suggests that depending on the required application, winding layout varies. In terms of copper loss, fault tolerance, and reliability, FSCW is an absolute choice. On the other hand, ISDW is capable to develop a higher torque.

Electric motors used in EVs must maintain the efficiency in a wide range of speed. Performance of four kinds of SRM, IM, concentrated winding IPM, and distributed winding IPM has been compared by Yang et al. (2015). It has been shown that IM and SRM have high efficiency only when they operate in a narrow-speed region, while the operation regions of two IPM motors are much wider. An optimization scheme for "extended-speed" region has been proposed by Zhang et al. (2016) where the characteristic current (Eqs. 2 and 3) has been increased in the presented IPM motor. A proposed variable leakage flux IPM motor has been considered in the work of Kato et al. (2015). It is noted that the low-load motor needs less flux linkage, so the flux linkage of PMs can be reduced by increasing the leakage flux. This leakage flux, controlled by the stator current, flows in a shorter magnetic path, and thus produces less core losses. This ensures that the high efficiency is maintained even over low-load region:

$$I_{ch} = \frac{\lambda_{pm}}{L_d} \tag{2}$$

$$(i_d + I_{ch})^2 + \left(\frac{L_q}{L_d}\right)^2 {i_q}^2 = \frac{{V_{\max}}^2}{{L_d}^2{}_e^2} \tag{3}$$

The short-duty capability of electric motor as an important aspect of efficiency has been discussed by Deshpande et al. (2015). In some applications such as lightweight urban EVs, this is a prominent matter. The outer rotor SPM motor for in-wheel application has been analyzed and importance of heat exchanging between the copper and the stator core has been notified. The copper bars have been suggested which enhance the heat conduction.

5 Flux Switching Permanent Magnet Motors

For high-speed operation, flux switching permanent magnet (FSPM) motor is a potentially viable solution. There are certain constraints in IPM motors and FSPM motors do not have such constraints. As a general rule, mechanical and magnetic properties of IPM motors are contradicting. The bridges between the PMs should have enough thickness that maintains mechanical robustness and also they must be as thin as possible in order to minimize the leakage flux. Topology of FSPM motor has no such limitations, because both PM and armature winding have been located on the stator.

A flux weakening method can be applied to the FSPM motor (Deodhar et al. 2014). The conventional flux weakening method normally uses d-component of the current to control the flux linkage, which causes excessive copper loss. To overcome this problem a mechanical approach has been proposed by Deodhar et al. (2014). Since the motor has a flux switching structure, all active parts are located on the stator and mechanical methods are much easier to apply. The underlying principle is simple: a set of flux diverting iron segments are embedded on the outer periphery of the stator. These diverters start to get closer to the stator in high speeds and shorten the magnetic circuit.

A torque pulsation optimization can be conducted on a high-efficiency FSPM motor in in-wheel application (Fei et al. 2012). It has been mentioned that cooling procedure of FSPMs is a much simpler task since there are no active parts on the rotor. The IPM and FSPM motors have been compared by McFarland et al. (2015) and it has been shown that the latter is a better choice for low-cost magnets in terms of demagnetization. Usage of similar structures in which wound-field excitation has been adopted instead of PMs has been proposed by Nguyen et al. (2016) and Raminosoa et al. (2015). Although they have lower efficiency, the stator is more robust and flux-weakening process is enhanced. In addition, there are in-between structures (Sulaiman et al. 2011), which is a hybrid structure with both PM and wound-field excitations. These motors have the merits of both FS structures.

6 Comparison and Possible Future Applications

Considering the latest advancements in electric motors, new applications are possible for certain types of motors. There are some applications for new electric motors that were not imaginable before. Utilization of linear FSPM electric motor in transportation is one of these newly realized applications. Prior to the development of the linear FSPM electric motors, the rotating PMS motors were the only option for linear movement. In the linear PMS motor, the moving part can be either primary or secondary, but in railway traction applications none of these two options are cost effective. With the emersion of the FSPM motor, this was no longer an issue. Since both armature and excitation field are placed on the stator (primary), it is highly desirable in traction application. A comparison between the FSPM and PMS motors has been conducted by Tang et al. (2012) for in-wheel application. It has been reported that for an equal volume, the peak torque of the FSPM motor is higher than that of the PMS motor. Flux-weakening capability is also higher in the latter motor.

The structures of rotor of flux switching motor (FSM) and conventional SRM are similar, and they can be compared in terms of acoustic noise (Pollock and Brackley 2001). It has been concluded by Pollock and Brackley (2001) that under the same conditions, the noise level of FSM is about 2 dB lower than that of the SRM. However, it is noted that the SSRM as an alternative version of SRM has less noise than that of the conventional SRM. This means that now they can be used in the lightweight and urban electric vehicles, in addition to their previous applications in heavy machinery such as loaders (Jahns 2017). Another noise-sensitive application has been introduced by Pollock et al. (2003) in which FSM has been compared with an IM for driving fan. It has been observed that FSM structure is a more efficient, but noisier option.

IM is dominant in almost every application, but more efficient designs make them comparable to PM electric motors. The line-start PMSM (LSPMSM) and IM have been compared by Pollock et al. (2003). In the steady-state mode, the LSPMSM is superior in every aspect including efficiency, power factor, and full load current but in startup transients, IM develops larger torque and has a better dynamic behavior and smooth movement.

7 Conclusion

This chapter reviews the latest advancements of a variety of electric motors. Many different techniques and methods were proposed which directly or indirectly affect the efficiency of electric motors. Due to the multidisciplinary nature of the electric motors, a wide variety of fields are involved in the development of these technologies. So, these developments were reported and categorized by their origins. Based on the latest advancements of electric motors, possible future applications of a few types of motors were discussed. Furthermore, different types of motors in different applications were compared and their weaknesses and strengths were mentioned. A more in-depth study of each structure can be presented in the future.

References

Alberti L, Bianchi N, Boglietti A, Cavagnino A (2014) Core axial lengthening as effective solution to improve the induction motor efficiency classes. IEEE Trans Ind Appl 50(1):218–225

de Almeida AT, Ferreira FJ, Fong JA (2009) Standards for Super-Premium Efficiency class for electric motors. In: Industrial & Commercial Power Systems Technical Conference-Conference Record 2009 IEEE, pp 1–8

Deodhar RP, Pride A, Iwasaki S, Bremner JJ (2014) Performance improvement in flux-switching PM machines using flux diverters. IEEE Trans Ind Appl 50(2):973–978

Deshpande YB, Toliyat HA, Nair SS, Dhinagar SJ, Immadisetty S, Nalakath S (2015) High-torque-density single tooth-wound bar conductor permanent-magnet motor for electric two wheeler application. IEEE Trans Ind Appl 51(3):2123–2135

Dorrell DG (2014) The challenges of meeting IE4 efficiency standards for induction and other machines. In: Industrial Technology (ICIT), 2014 IEEE International Conference on 2014, pp 213–218

Fan T, Li Q, Wen X (2014) Development of a high power density motor made of amorphous alloy cores. IEEE Trans Ind Electron 61(9):4510–4518

Fei W, Luk PCK, Shen JX, Wang Y, Jin M (2012) A novel permanent-magnet flux switching machine with an outer-rotor configuration for in-wheel light traction applications. IEEE Trans Ind Appl 48(5):1496–1506

Finley WR, Hodowanec MM (2001) Selection of copper versus aluminum rotors for induction motors. IEEE Trans Ind Appl 37(6):1563–1573

Gonzalez A, Hernandez C, Arjona M (2014) A novel high-efficiency parallel-winding connection for a three-phase induction motor fed by a single-phase power supply. IEEE Trans Energy Conv 29(2):269–277

Goss J, Popescu M, Staton D (2013) A comparison of an interior permanent magnet and copper rotor induction motor in a hybrid electric vehicle application. In: Electric Machines \& Drives Conference (IEMDC), 2013 IEEE International, pp 220–225

Jahns T (2017) Getting rare-earth magnets out of EV traction machines: a review of the many approaches being pursued to minimize or eliminate rare-earth magnets from future EV drivetrains. IEEE Electrification Mag 5(1):6–18

Jang K (2017) Status of high-efficiency motor technology and MEPS. J Electr Eng Technol 12(2):803–808

Kato T, Minowa M, Hijikata H, Akatsu K, Lorenz RD (2015) Design methodology for variable leakage flux IPM for automobile traction drives. IEEE Trans Ind Appl 51(5):3811–3821

Kim K-C, Koo D-H, Hong J-P, Lee J (2007) A study on the characteristics due to pole-arc to pole-pitch ratio and saliency to improve torque performance of IPMSM. IEEE Trans Magn 43(6):2516–2518

Kim K-C, Lee J, Kim HJ, Koo D-H (2009) Multiobjective optimal design for interior permanent magnet synchronous motor. IEEE Trans Magn 45(3):1780–1783

Kiyota K, Kakishima T, Chiba A, Rahman MA (2016) Cylindrical rotor design for acoustic noise and windage loss reduction in switched reluctance motor for HEV applications. IEEE Trans Ind Appl 52(1):154–162

Lee G-H, Kim S-I, Hong J-P, Bahn J-H (2008) Torque ripple reduction of interior permanent magnet synchronous motor using harmonic injected current. IEEE Trans Magn 44(6):1582–1585

Li L, Zhang J, Zhang C, Yu J (2016) Research on electromagnetic and thermal issue of high-efficiency and high-power-density outer-rotor motor. IEEE Trans Appl Supercond 26(4):1–5

Li C, Xu D, Wang G (2017) High efficiency remanufacturing of induction motors with interior permanent-magnet rotors and synchronous-reluctance rotors. In: Transportation electrification Asia-Pacific (ITEC Asia-Pacific), 2017 IEEE conference and Exp, pp 1–6

Lin C-H, Hwang C-C (2016) Multiobjective optimization design for a six-phase copper rotor induction motor mounted with a scroll compressor. IEEE Trans Magn 52(7):1–4

Liu G, Zeng Y, Zhao W, Ji J (2018) Permanent magnet shape using analytical feedback function for torque improvement. IEEE Trans Ind Electron 65(6):4619–4630

Lu S-M (2016) A review of high-efficiency motors: specification, policy, and technology. Renew Sust Energ Rev 59:1–12

Madhavan R, Fernandes BG (2014) Performance improvement in the axial flux-segmented rotor-switched reluctance motor. IEEE Trans Energy Conv 29(3):641–651

Malinowski J, McCormick J, Dunn K (2004) Advances in construction techniques of AC induction motors: preparation for super-premium efficiency levels. IEEE Trans Ind Appl 40(6):1665–1670

McFarland JD, Jahns T, El AM (2015) Performance and efficiency comparisons for interior PM and flux-switching PM machines with ferrite magnets for automotive traction applications. In: Energy Conversion Congress and Exposition (ECCE), 2015 IEEE, pp 6529–6536

Mecrow B, Finch J, El-Kharashi E, Jack A (2002) Switched reluctance motors with segmental rotors. IEE Proceedings-Electric Power Applications 149(4):245–254

Mecrow BC, El-Kharashi EA, Finch JW, Jack AG (2004) Preliminary performance evaluation of switched reluctance motors with segmental rotors. IEEE Trans Energy Conv 19(4):679–686

Min SG, Bramerdorfer G, Sarlioglu B (2018) Analytical modeling and optimization for electromagnetic performances of fractional-slot PM brushless machines. IEEE Trans Ind Electron 65(5):4017–4027

Mousavi-Aghdam S, Feyzi M, Bianchi N (2017) Analysis and comparison study of novel stator-segmented switched reluctance motor. Iran J Electr Electron Eng 13(1):68

Nguyen HQ, Jiang J-Y, Yang S-M (2016) Design of a 12-slot 7-pole wound-field flux switching motor for traction applications. In: Industrial Technology (ICIT), 2016 IEEE International Conference on 2016, pp 1275–1280

Ni R, Xu D, Wang G, Gui X, Zhang G, Zhan H, Li C (2016) Efficiency enhancement of general AC drive system by remanufacturing induction motor with interior permanent-magnet rotor. IEEE Trans Ind Electron 63(2):808–820

Oyama J, Higuchi T, Abe T, Kifuji N (2006) Novel switched reluctance motor with segment core embedded in aluminum rotor block. IEEJ Trans Ind Appl 126(4):385–390

Pollock C, Brackley M (2001) Comparison of the acoustic noise of a flux switching and a switched reluctance drive. In: Industry applications conference, 2001. Thirty-sixth IAS annual meeting. Conference Record of the 2001 IEEE, vol 3, pp 2089–2096

Pollock C, Pollock H, Brackley M (2003) Electronically controlled flux switching motors: a comparison with an induction motor driving an axial fan. In: Industrial Electronics Society, 2003. IECON'03. The 29th Annual Conference of the IEEE, vol 3, pp 2465–2470

Rajkumar S, SedhuRaman K, Balaji D, Kannan S (2017) High torque to weight ratio and cost estimation of cage induction motor with finite element method for hybrid vehicle. In: Innovative Research in Electrical Sciences (IICIRES), 2017 International Conference on 2017, pp 1–5

Raminosoa T, El-Refaie AM, Pan D, Huh K-K, Alexander JP, Grace K, Grubic S, Galioto S, Reddy PB, Shen X (2015) Reduced rare-earth flux-switching machines for traction applications. IEEE Trans Ind Appl 51(4):2959–2971

Spargo CM, Mecrow BC, Widmer JD, Morton C (2015) Application of fractional-slot concentrated windings to synchronous reluctance motors. IEEE Trans Ind Appl 51(2):1446–1455

Sulaiman E, Kosaka T, Matsui N (2011) High power density design of 6-slot-8-pole hybrid excitation flux switching machine for hybrid electric vehicles. IEEE Trans Magn 47(10):4453–4456

Tang Y, Motoasca E, Paulides JJ, Lomonova EA (2012) Comparison of flux-switching machines and permanent magnet synchronous machines in an in-wheel traction application. Int J Comput Math Electr Electron Eng 32(1):153–165

Tong W, Wu S, Sun J, Zhu L (2016) Iron loss analysis of permanent magnet synchronous motor with an amorphous stator core. In: Vehicle Power and Propulsion Conference (VPPC), 2016 IEEE, pp 1–6

Tsunata R, Takemoto M, Ogasawara S, Watanabe A, Ueno T, Yamada K (2017) Development and evaluation of an axial gap motor using neodymium bonded magnet. IEEE Transactions on Industry Applications

Vandana R, Fernandes B (2015) Design methodology for high-performance segmented rotor switched reluctance motors. IEEE Trans Energy Conv 30(1):11–21

Vattikuti N, Rallabandi V, Fernandes B (2008) A novel high torque and low weight segmented switched reluctance motor. In: Power electronics specialists conference, 2008. PESC 2008. IEEE, pp 1223–1228

Wang X, Zhong H, Yang Y, Mu X (2010) Study of a novel energy efficient single-phase induction motor with three series-connected windings and two capacitors. IEEE Trans Energy Conv 25(2):433–440

Wang K, Zhu Z, Ombach G, Chlebosz W (2014) Average torque improvement of interior permanent-magnet machine using third harmonic in rotor shape. IEEE Trans Ind Electron 61(9):5047–5057

Widmer JD, Martin R, Mecrow BC (2015) Optimization of an 80-kW segmental rotor switched reluctance machine for automotive traction. IEEE Trans Ind Appl 51(4):2990–2999

Xu L, Liu G, Zhao W, Ji J, Zhou H, Zhao W, Jiang T (2015) Quantitative comparison of integral and fractional slot permanent magnet Vernier motors. IEEE Trans Energy Conv 30(4):1483–1495

Xu Z, Kim M-J, Lee D-H, Ahn J-W (2016) Characteristics analysis and comparison of conventional and segmental rotor type 12/8 switched reluctance motors. In: Industry Applications Society Annual Meeting, 2016 IEEE, pp 1–7

Yang Z, Shang F, Brown IP, Krishnamurthy M (2015) Comparative study of interior permanent magnet, induction, and switched reluctance motor drives for EV and HEV applications. IEEE Trans Transport Electrification 1(3):245–254

Yoon K-Y, Kwon B-I (2016) Optimal design of a new interior permanent magnet motor using a flared-shape arrangement of ferrite magnets. IEEE Trans Magn 52(7):1–4

Zhang L, Huang Y, Dong J, Guo B, Zhou T (2014) Stator winding design of induction motors for high efficiency. In: Electrical Machines and Systems (ICEMS), 2014 17th International Conference on 2014, pp 130–134

Zhang P, Ionel DM, Demerdash NA (2016) Saliency ratio and power factor of IPM motors with distributed windings optimally designed for high efficiency and low-cost applications. IEEE Trans Ind Appl 52(6):4730–4739

Zhang D, An R, He C, Bu L, Wu T (2017) Electromagnetic design of a megawatt high efficiency high speed solid rotor induction motor. In: Electric Machines and Drives Conference (IEMDC), 2017 IEEE International, pp 1–8

Zhao W, Lipo TA, Kwon B-I (2014a) Comparative study on novel dual stator radial flux and axial flux permanent magnet motors with ferrite magnets for traction application. IEEE Trans Magn 50(11):1–4

Zhao W, Lipo TA, Kwon B-I (2014b) Material-efficient permanent-magnet shape for torque pulsation minimization in SPM motors for automotive applications. IEEE Trans Ind Electron 61(10):5779–5787

Zhao W, Zhao F, Lipo TA, Kwon B-I (2014c) Optimal design of a novel V-type interior permanent magnet motor with assisted barriers for the improvement of torque characteristics. IEEE Trans Magn 50(11):1–4

Zhao W, Lipo TA, Kwon B-I (2015a) A novel dual-rotor, axial field, fault-tolerant flux-switching permanent magnet machine with high-torque performance. IEEE Trans Magn 51(11):1–4

Zhao W, Lipo TA, Kwon B-I (2015b) Dual-stator two-phase permanent magnet machines with phase-group concentrated-coil windings for torque enhancement. IEEE Trans Magn 51(11):1–4

Zhao W, Lipo TA, Kwon B-I (2015c) Torque pulsation minimization in spoke-type interior permanent magnet motors with skewing and sinusoidal permanent magnet configurations. IEEE Trans Magn 51(11):1–4

Zhao W, Chen D, Lipo TA, Kwon B-I (2017a) Dual airgap stator-and rotor-permanent magnet machines with spoke-type configurations using phase-group concentrated coil windings. IEEE Trans Ind Appl 53(4):3327–3335

Zhao W, Kwon J, Wang X, Lipo TA, Kwon B (2017b) Optimal design of a spoke-type permanent magnet motor with phase-group concentrated-coil windings to minimize torque pulsations. IEEE Trans Magn 53(6):1–4

Zhao W, Xing F, Wang X, Lipo TA, Kwon B (2017c) Design and analysis of a novel PM-assisted synchronous reluctance machine with axially integrated magnets by the finite-element method. IEEE Trans Magn 53(6):1–4

Predictability vs. Efficiency of Large-Scale Multi-Agent Systems

Predrag T. Tošić

Abstract We want to characterize "(in)efficient" and "(un)predictable" dynamics of large-scale *multi-agent systems* (MAS). We abstract the collective behaviors of such MAS as appropriate classes of cellular and network automata, and define *computational notions* of predictability and efficiency of those discrete networks' dynamics. We say that a discrete network's dynamics is efficient if it settles quickly into an appropriate stationary pattern, such as a stable ("fixed point") configuration or a temporal cycle; that is, efficiency for us is synonymous with a short transient chain until some stationary behavior is reached.

The (in)efficiency of a deterministic networked dynamical system is related to, but not synonymous with, that network's (un)predictability. We introduce two computational notions of predictability: *local* and *global*. We call network dynamics locally predictable; if given a starting configuration, the convergence to a stationary behavior can be efficiently determined or predicted, without performing a step-by-step simulation—unless such simulation itself can be done efficiently. A network's dynamics is globally predictable, if we can computationally efficiently characterize all possible dynamics for all initial configurations of the system. By *efficiently* throughout the paper, we mean in the time at most polynomial in the number of network's nodes.

For discrete networks with deterministic local interactions and finite configuration spaces, all properties of such systems' dynamics are formally decidable; however, for nontrivial network sizes, particular properties may or may not be possible to determine within reasonable computational resources. We overview computational complexity of the reachability-flavored problems as well the problems of enumeration of different dynamics of a given dynamical system across all possible initial configurations, studied in the context of several classes of cellular automata, Boolean networks, and discrete Hopfield networks. We argue

P. T. Tošić (✉)
Independent Researcher & Scholar, Spokane, WA, USA

Division Department of Mathematics and Statistics (Adjoint Research Faculty), Washington State University, Pullman, WA, USA
e-mail: predrag.tosic@ieee.org

G. Y. Georgiev, M. Shokrollahi-Far (eds.), *Efficiency in Complex Systems*,
Springer Proceedings in Complexity, https://doi.org/10.1007/978-3-030-69288-9_6

that the reachability problems about those systems' asymptotic dynamics have to be computationally tractable, in order for a given system to have globally predictable dynamics. We also relate the problem of enumerating all possible dynamical evolutions of a network to its global (un)predictability.

We study some interesting examples of cellular automata and Boolean networks whose dynamics can be shown to be (i) both predictable and efficient, (ii) efficient but unpredictable, (iii) predictable yet inefficient, and last but not least, (iv) both unpredictable and inefficient. Lastly, we briefly discuss Boolean networks that are globally unpredictable, yet whose efficiency status is still open.

Keywords Cellular automata · Boolean networks · Discrete Hopfield networks · Asymptotic dynamics · Computational complexity of enumeration problems · Computational complexity of reachability problems · Efficiency of system's dynamics · Predictability of system's dynamics

1 Introduction and Motivation

We have been investigating abstract models of large-scale *multi-agent systems* (MAS) as deterministic discrete networks (Tosic 2005, 2006a, 2017). We study collective dynamics and emerging behaviors of such MAS by analyzing the underlying discrete networks' formal dynamics, that is, their configuration spaces. We are particularly interested in those networks' long-term, that is, *asymptotic dynamics* (Barrett et al. 2003; Tosic 2005, 2017; Tosic and Ordonez 2017). In this chapter, we focus on (deterministic, discrete time) Boolean networks, in which each node can be in one of only two distinct states. Such models have been called *Boolean networks* (BNs) or *Boolean network automata* (BNA) in the existing literature, and they represent a generalization of the classical *cellular automata* (CA) that have been studied for over half a century.

Boolean networks have been extensively used as a modeling tool in statistical physics, theoretical and systems biology, computational neuroscience, as well as network science and agent-based modeling of a broad variety of cyber-physical, cyber-secure, socio-technical, and other large-scale decentralized networked systems. In these models, each node is a simple deterministic *finite-state machine* (FSM), where the behavior of each node is specified as some function of (usually) its own current state and the states of some subset of its neighboring nodes (Tosic and Agha 2004a; Tosic 2010a, b).

Classical *cellular automata* (CA) are perhaps the best known example of such discrete Boolean networks. A cellular automaton is defined over a regular Cayley graph such as a line or ring (1-dimensional cellular spaces) or a higher dimensional grid or torus (for 2- and higher dimensional cellular spaces). Each node in a CA updates its state according to the same update rule—some function of (usually) the current state of that node itself, and the states of (in general, some subset of) its neighbors (Tosic and Agha 2004a, b; Wolfram 1984, 1986, 1994). CA generalize to various other network automata models, in which (i) the network topology can be more general than the highly regular cellular spaces of CA, and (ii) different nodes are allowed to update their states according to different update rules.

There have been a number of such generalizations of the basic CA model. One well-known class of such models is the (discrete) *Hopfield networks* (Hopfield 1982). Hopfield networks have been used since the 1980s to model various phenomena mainly in statistical physics and theoretical biology (Hopfield 1982; Hopfield and Tank 1985). More recently, Hopfield networks have been studied as an abstract model of computational brains approximated as *associative memories*; see, e.g., Davey et al. (2006) and Tosic (2016).

Discrete network models have been used in other theoretical biology contexts, as well—such as for mathematical modeling related to protein networks and gene regulatory networks; see, e.g., Graudenzi et al. (2011) and Kauffman (1984). Yet other, descriptively fairly similar, discrete network models have been studied in the context of modeling various cyber-physical, socio-technical, and multi-agent systems (Bandini et al. 2002; Tosic 2005, 2006a, 2011, 2017). Our own interest in these models originated from two directions: theory of concurrency and communication models for the large-scale distributed computing on the one hand (Tosic and Agha 2004a, b; Tosic 2010a), and formal models for distributed AI and multi-agent systems, on the other (Tosic 2005, 2010b, 2017, 2002).

Among many interesting questions about the dynamics of these various discrete network models, their *asymptotic dynamics*, determining what will the system do in the long run starting from either an arbitrary or a certain set of specific initial configuration(s), have been of a major interest. Characterizing in how many different ways can such a network evolve, and, in particular, enumerating (exactly or at least approximately) all of its stable configurations ("fixed points") have been studied in the context of Hopfield networks since the 1990s (Floreen and Orponen 1989; Orponen 1996; Sima and Orponen 2003). More recently, our extension of that line of research has resulted in tighter theoretical results on the complexity of counting Hopfield networks' stable configurations, as well as in some interesting interpretations related to the storage capacity of associative memories (Tosic 2016).

Enumeration of possible dynamics in general, and of stable ("fixed point") configurations in particular, has been investigated in the context of discrete network models of large-scale multi-agent and cyber-physical systems, as well (Tosic and Agha 2005; Tosic 2006a, 2010b, 2017). In that area of research, we have made a considerable progress in attempting to characterize the "phase transitions" from relatively simple and predictable network dynamics to provably complex and unpredictable dynamics, as the network parameters such as the underlying network topology and the local update rules are "tweaked" in various ways (Tosic 2010b, 2017). In particular, the present work, focusing on the computational notions of *efficiency* and *predictability*, can be considered a continuation of our research on phase transitions in network dynamics' complexity found in Tosic (2017) and Tosic and Ordonez (2017).

We define several discrete network models, previously used in applications as broad as statistical physics, theoretical biology, multi-agent, and networked cyber-physical systems; in the next section, we will restrict our attention to *Boolean networks*. We are interested in the efficiency and predictability of the asymptotic dynamics of these discrete-time Boolean networks. We note that both efficiency and predictability can be defined in many different ways. For example, efficiency of a system can be measured in terms of the energy flow and dissipation, information flow into and out of a system, and so on. However, we are interested in intrinsically computational definitions of the concepts of a system's dynamics' efficiency and predictability.

Since we restrict our discussion to finite, fully deterministic discrete dynamical systems, all properties of such systems' dynamics can always be determined "in principle"; that is, they are formally decidable. The problems of efficiency and predictability, then, reduce to which aspects about the system dynamics pertaining to how efficient and how predictable that dynamics is can be determined *computationally efficiently*, and which aspects would take prohibitively large computational resources for nontrivial network sizes. For simplicity, we will consider the number of nodes of a discrete network to be the parameter that, for our purposes, adequately captures "the network size." Denoting that number of nodes with n, it is immediate that the size of the entire configuration space of a Boolean network with n nodes is 2^n.

2 Definitions, Preliminaries, and Prior Art

In this section, we define several discrete network models of interest; in their Boolean variants, each node in the underlying graph is in one of the two possible states, 0 or 1. We then formulate some fundamental properties of *the dynamics* of *Boolean networks* and, as their prominent special cases, classical *cellular automata* (Tosic and Agha 2004a; Wolfram 1984, 1986, 1994) as well as *discrete Hopfield networks* (Floreen and Orponen 1989; Hopfield 1982; Hopfield and Tank 1985; Tosic 2016).

Definition 1 A *Boolean network* is a directed or undirected graph so that each node in the graph has a state, 0 or 1. Each node periodically updates its state, as a function of the current states of its neighboring nodes.

Discrete Boolean networks are also called *Boolean network automata* (BNA for short) in the literature. We observe that the local update rule being a function of the neighboring nodes should be interpreted loosely, namely, such update rule is a function of some (in particular, possibly, but not necessarily, all) of its neighboring nodes' states; likewise, a node's update rule possibly, but not necessarily, depends on the current state of that node itself.

A BN dynamically evolves in discrete time steps. If the node v_i has k neighbors (in a loose sense as outlined above), denoted as $v_{i1}, \ldots, v_{ik}$, then the next state of vi is determined by evaluating a Boolean-valued function $f_i(v_{i1}, \ldots v_{ik})$ of k Boolean variables; f_i is called the *local update function* or *transition rule* for the node v_i. We restrict our attention to the models where each node's next state depends on its own current state (these are the so-called *automata with memory*), although we point out that all our results in this chapter hold irrespective of this particular assumption.

Some clarifications are due. First, in network automata models more general than the classical cellular automata, different nodes v_i may use different local update rules f_i. This applies to *discrete Hopfield networks* (Davey et al. 2006; Floreen and Orponen 1989; Hopfield 1982; Hopfield and Tank 1985), as well as many other classes of Boolean networks, such as those originally introduced by S. Kauffman in the context of systems biology (Graudenzi et al. 2011; Kauffman 1984), and also several loosely related models proposed in the context of modeling large-scale decentralized cyber-physical, multi-agent, and socio-technical systems and infrastructure; see, e.g., Barrett et al. (2001), Barrett et al. (2003), Mortveit and Reidys (2001), and Tosic (2006b). Classical *cellular automata* (CA), on the other hand, can be viewed as a special case of BNs, where all the nodes use the same local update rule f_i (Tosic and Agha 2004a; Tosic 2010b; Wolfram 1994).

The individual node updates can be done either perfectly synchronously in parallel or sequentially, one at a time. While other communication models are worth considering (see, e.g., Tosic (2010a) and Tosic (2011)), the above two possibilities have been studied the most. In this chapter, we will focus for the most part on the parallel, *perfectly synchronous* node updates (Tosic 2010a). That is, the next state of the node v_i is determined according to

$$v_i^{t+1} \leftarrow f_i\left(v_{i1}^{t}, \ldots, v_{ik}^{t}\right) \tag{1}$$

where the superscript t denotes a discrete time step.

The tuple of all f_i's put together, $F = (f_1, \ldots, f_n)$, denotes the *global map* that acts on global configurations of a Boolean network. When all f_i are the same, as is the case with classical cellular automata, the notation is commonly abused so that no differentiation is made between the local transition function f_i, acting on a state of a single node, and the global map F that acts on all the nodes.

As we note, in most Boolean network models other than the classical CA, different nodes in general update according to different local rules. This also applies to Hopfield networks, defined next.

Definition 2 A *discrete Hopfield network* (DHN) is made of n binary-valued nodes. Associated to each pair of nodes (v_i, v_j) is their (in general, real-valued) weight, w_{ij}. The weight matrix of a DHN is defined as $W = [w_{ij}]_{i,j=1..n}$. Each node also has a fixed real-valued threshold, h_i. A node v_i updates its state x_i from time step t to step $t + 1$ according to a (binary-valued) *linear threshold function* of the form:

$$x_i^{t+1} \leftarrow \mathrm{sgn}\left(\sum_j w_{ij} \bullet x_j^{t} - h_i\right) \tag{2}$$

where the summation is over $j = 1, \ldots, n$; the term h_i is the *threshold* that the weighted sum needs to exceed in order for the node's state to update to +1; to break ties, we define sgn(0) = +1.

For consistency across all discrete network models we discuss in this work, we will always denote the set of states of a single node as {0, 1} and apply that same convention to Hopfield networks (for which, in most of the existing literature, the two "default" state labels are {−1, +1}). A bijection R ➔ (1 + R)/2 achieves the desired mapping from the set of states {−1, +1} to {0, 1}, where R stands for the right-hand side in Eq. (2) above.

We summarize insights about efficiency and predictability of various kinds of Boolean networks based on their asymptotic dynamics. Some of our results were originally formulated in the DHN context (Tosic 2016); yet other results were formulated in the context of two other types of network automata (whose nodes' states in general come from a finite set and, in particular, need not necessarily be Boolean valued), called *sequential and synchronous dynamical systems* (Barrett et al. 2001, 2003*;* Tosic and Agha 2005*;* Tosic 2005, 2010b, 2017).

Hopfield networks (both those that evolve discreetly and those that evolve continuously in time) were originally inspired by statistical physics, systems biology, and computational neuroscience (Hopfield 1982). Subsequently, discrete and continuous Hopfield networks have also been used as a connectionist, self-organizing map-based technique for "learning" and "searching for a solution," i.e., as a powerful tool for various search and optimization problems in computer science and operations research (Hopfield and Tank 1985).

Some of the earliest Boolean network models other than Hopfield nets were also originally introduced in the context of theoretical and systems biology, although not specifically neuroscience. Indeed, the very name *Boolean networks* comes from the seminal work in theoretical biology by S. Kauffman (Kauffman 1984). In contrast, *sequential* and *synchronous dynamical systems* (SDS and SyDS, resp.) were introduced in the context of agent-based simulation of complex cyber-physical, socio-technical, and physical systems (Barrett et al. 2003; Mortveit and Reidys 2001; Tosic and Agha 2005; Tosic 2005, 2006a). We do not formally define these S(y)DS models here; instead, we refer the reader to references (Barrett et al. 2001, 2003; Mortveit and Reidys 2001; Tosic and Agha 2005; Tosic 2006a, b).

We note that all the results that will be discussed in this chapter, and the implications of those results for the discrete dynamical system's efficiency and predictability, apply to all of the above models (CA, DHNs, SDSs, and SyDSs), and indeed other similar discrete-time Boolean (or other finite-domain) networks found in the literature.

Since cellular automata, Boolean networks, and DHNs are all deterministic discrete-time dynamical systems, for any given current configuration C^t at time t, there is a unique next-step configuration C^{t+1}. We can therefore define the BN or DHN *configuration spaces*, and also various types of *global configurations* of interest:

Definition 3 A (global) configuration of a cellular or network automaton or a discrete Hopfield network is a vector $(x_1, \ldots, x_n) \in \{0,1\}^n$, where x_i denotes the current state of the ith node.

A global configuration can also be thought of as a function $F: V \rightarrow \{0,1\}$, where V denotes the set of nodes in the underlying graph of a CA, BN, or DHN.

Definition 4

- A *fixed point* (FP) is a global configuration such that, once a CA, BN, or DHN reaches that configuration, it stays there forever.
- A *cycle configuration* (CC) is a global configuration that, once reached, will be revisited infinitely often with a fixed, finite temporal period of 2 or greater.
- A *transient configuration* (TC) is a global configuration that, once reached, is never going to be revisited again. A *Garden-of-Eden* (GE) configuration is a TC that has no predecessors; that is, GEs can only occur as the initial configurations of the system.

How do we relate *predictability* of a discrete network to its asymptotic dynamics? We call a deterministic dynamical system predictable, if given an arbitrary starting configuration we can predict the system's "ultimate destiny," without necessarily having to run a step-by-step simulation. What can that long-term behavior be?

For the *deterministic* discrete models of our interest such as the classical CA, the more general Boolean networks, and discrete Hopfield networks, eventually the system behavior will become stationary. That is, one of the following two things will happen: a FP configuration will be reached, or else, a temporal cycle (of some fixed finite length greater than or equal to 2) will be reached. The underlying system's dynamics is then predictable, if we can tell *which* FP or temporal cycle will be reached—and if we can do so in a computationally efficient manner, that is, using a computation that takes a number of steps polynomial in the number of network nodes, *n*.

For certain dynamical systems, one can determine that "ultimate destiny" analytically; we will briefly discuss some such examples in the next section. For others, simulating the system step by step might be acceptable, if such convergence to a FP or a temporal cycle takes a number of steps *polynomial in the number of nodes* (which, in particular, means that it is *polylogarithmic* in the size of the configuration space, 2^n). For yet others, there may not be a computationally feasible way of determining the asymptotic behavior without a step-by-step simulation and, moreover, such simulation may take a number of steps that is exponential in *n*. The first two types of systems, according to our definition, would then be considered predictable, whereas the third kind would be considered unpredictable.

Due to the deterministic nature of these interactions and finiteness of configuration spaces involved, it is always possible to determine the asymptotic dynamics (or, for that matter, any other aspect of network dynamics) by simulating the system step by step, starting if necessary from each of the possible 2^n global configurations. The predictability then, from our computational standpoint, reduces to when can these questions about the asymptotic dynamics be answered (analytically or via simulation) in a number of steps that is only polynomial, as opposed to exponential, in the number of network nodes *n*.

Definition 5 The *reachability* problem for deterministic discrete dynamical systems is, given two global configurations C and C', to determine whether, starting at the state C, the system will ever reach the state C'.

Definition 5′ The *FP reachability* problem is formulated as follows: starting from an arbitrary configuration C, what will be the ultimate destiny of that system's dynamics, in particular, whether a FP will eventually be reached?

Definition 5″ The *temporal cycle reachability* problem asks, starting from an arbitrary initial configuration C, whether a temporal cycle (of length 2 or greater) will eventually be reached?

For many classes of these deterministic dynamical systems, even determining whether the "ultimate destiny" of the system (starting from a given global state) will be a fixed point or a temporal cycle is hard to determine. A refined version of this problem is whether we can efficiently determine, in case of FP reachability, *which* particular FP will be eventually reached, and similarly for the temporal cycle reachability problem. It has been established that, for broad classes of discrete networks, the reachability problem in general tends to be **PSPACE**-complete (Barrett et al. 2003); this means that, under the usual assumptions in the computational complexity theory, the reachability questions cannot be answered in time polynomial in the size of the system. Further, there are interesting classes of CA and BNs, some of which we shall discuss in the next section, for which the FP reachability question is trivial to answer, simply because the underlying dynamical system *does not have any temporal cycles*—or, in some cases, even if it does, as is the case for example with the *Majority* CA with parallel node updates, those CCs do not have any incoming transients (Tosic and Agha 2004a). Hence, for all (or almost all) initial states, *we know that such system will end up in (some) fixed point.* However, in that case, the more refined FP reachability question, namely to which particular one among possibly exponentially many FPs will the system converge starting from a given initial configuration, often remains intractable. We discuss some examples of such BNs and DHNs in Sect. 4.

On the other hand, for certain restricted classes of Boolean networks, as well as several important classes of (finite-sized) CA and Hopfield networks, the reachability questions, including the more specific FP reachability problem, can be answered efficiently, that is, in time that is polynomial in the network size. In particular, if the FP reachability can be determined in polynomial time for all possible starting configurations, we say that the dynamics of such system is (globally) predictable. Moreover, if we can establish such predictability from a certain subset of candidate initial states, but perhaps do not know if the property holds starting from all possible initial states, we call such dynamical systems *locally predictable.*

Definition 6

- A deterministic Boolean network or other discrete dynamical system is *locally predictable* (from a particular initial state, or some set of initial states) if we can determine computationally efficiently what will be the asymptotic behavior of the system, starting from the designated initial state(s).
- A deterministic dynamical system is *globally predictable* if we can characterize all its possible asymptotic dynamics computationally efficiently, that is, in time at most polynomial in the size of the system's description.

In particular, global predictability trivially implies local predictability for *all* possible initial configurations.

This discussion of different types of predictability brings us to the second fundamental property of the discrete networks' dynamics that is of our interest: How *efficient* is that dynamics, in the worst case? In our context, we define a computational notion of efficiency to mean the following: starting from an arbitrary configuration, how quickly will the system settle down to its asymptotic, stationary behavior—whatever that ultimate behavior may be? That is, we do not care here, whether the ultimate destiny will be a FP or a temporal cycle, nor which particular FP or temporal cycle will it be. We only care about *how long it will take* to get there. Going back to the fundamental configuration space properties as defined earlier, this question can be rephrased as *how long is the transient chain prior to reaching the stationary state*.

Definition 7 A deterministic discrete dynamical system is *efficient*, if it reaches a stationary configuration (either an FP or a temporal cycle) in a number of global steps that is *at most polynomial* in the size of the system.

With these computational concepts of global predictability, local predictability, and efficiency, we start our analysis by establishing a relatively obvious preliminary result:

Theorem 1 *Efficiency implies local predictability*: If a deterministic discrete network's dynamics, starting from some initial configuration, is efficient, then it is also locally predictable.

Proof: Assume that this deterministic dynamical system has efficient dynamics starting from some initial configuration C. That means, the system converges to some stationary pattern within the number of steps that is at most polynomial in the number of nodes, *n*. Then we can predict the system's behavior, simply by running a step-by-step simulation starting from C, to see in which stationary behavior it will end up. This simulation will take at most polynomial time, i.e., it is computationally feasible, implying local predictability of the system's dynamics starting from that particular starting configuration.

While the above computational notions of network dynamics efficiency and predictability are related, and in particular efficiency implies local predictability as defined above, clearly *efficiency* and *global predictability* are not synonymous; further, as we shall argue in Sect. 4, neither property necessarily implies the other. So, which discrete networks have efficient vs. inefficient dynamics, and which networks' dynamics are predictable vs. unpredictable? We summarize what is known about a few popular classes of finite CA, Boolean networks, and discrete Hopfield networks in the next two sections.

3 Examples of Discrete Boolean Networks that Are both Predictable and Efficient

Given the above computational notions of dynamics' predictability and efficiency, what would be some examples of interesting classes of CA, Boolean networks, and Hopfield networks that are (globally) (un)predictable and/or (in)efficient? It turns out that there are interesting classes of such models in each of the four "quadrants" with respect to these two complexity dimensions. Certainly, a strong case could be made that those systems whose dynamics are predictable yet inefficient, and those whose dynamics are efficient yet unpredictable, are the most interesting ones.

In this section, we focus on the discrete networks that fall into the first quadrant, that is, whose dynamics are both globally predictable and efficient. In that context, we discuss examples of CA and BNs that either are already known to be or else can be analytically and/or computationally shown to be both efficient and predictable. We then summarize in the next section a few structurally rather simple BN and Hopfield net classes that turn out to be (globally) unpredictable in our sense; as we shall see, some of those dynamical systems are also inefficient, whereas others are efficient, and for some of them, the status of their efficiency is, to the best of our knowledge, yet to be determined.

We start with the classical CA with some finite number of nodes n. It has been known for over 30 years, based on the seminal work by S. Wolfram (Wolfram 1986, 1994) and others, that there are a number of local update rules that lead to highly complex and in general unpredictable behavior of even 1-dimensional CA, that is, a finite string or ring of nodes all of which update according to the same local rule. However, when the local update rule f_i is considerably restricted, such as to the linear threshold functions or the monotone functions, then usually one obtains the resulting dynamics that is both predictable and efficient. What are then some specific examples of such update rules resulting in predictable and efficient CA dynamics?

Consider a 1-dimensional CA in which each node updates according to the Boolean AND function; to how many inputs (that is, to how many nearby nodes to the left and to the right from the given node) this AND is applied to is determined by the rule radius r, where this radius is some integer $r > 0$. Typically, this means the input values to the local update rule are the r neighbors to the left, the current state of the node in question itself, and the r neighbors to the right, for the total of 2r + 1 input bits. It is not hard to establish that, for any starting configuration other than 1^n = (1, 1, . . . , 1), eventually the CA will converge to the FP "attractor" 0^n. Further, 1^n is also a FP; however, its entire "basin of attraction" is just that global configuration itself; that is, there are no incoming transients into 1^n. Likewise, either analytically or via (certifiably very efficient!) simulation, one can also readily establish that, starting from an arbitrary configuration $C \neq 1^n$, the convergence to the attractor 0^n will happen fast, and certainly in a number of steps polynomial in the number of nodes. (In fact, that speed of convergence is always O(n) and, within a constant multiplicative factor, the larger the radius r of the local update rule, the faster the convergence.)

The analysis for the CA in which each node updates according to Boolean OR is virtually identical to the analysis for AND, with the roles of states 0 and 1 reversed. In particular, OR CA only have two FPs, with 0^n being an isolated FP with no other predecessors but itself, and 1^n being the global attractor for all other 2^n-1 configurations. It then immediately follows that the CA in which all nodes update either according to the Boolean AND or according to the Boolean OR have asymptotic dynamics that is both efficient (the convergence is fast) and predictable (we can tell, to which FP will the system converge from any given starting configuration).

Are there update rules more interesting than Boolean AND and OR that also lead to provably efficient and predictable dynamics? It turns out that there are a number of other such node update functions, especially in the context of classical CA, in which all the nodes update their states according to the same local update rule. We discuss in some detail just one, perhaps most studied, such CA rule across a number of research disciplines and applications domains: the cellular automata whose nodes update according to the *majority* (MAJ) rule. For simplicity, we focus on the 1-D MAJ CA of radius r $\geq$1: node x_i updates its state to 1 if at least r +1 (out of 2r +1 input values total that include the current state of the node x_i itself) of its nearest neighbors are currently in the state 1, and the node updates its state to 0, otherwise.

How hard is to determine the dynamics of MAJ CA? In particular, for various underlying cellular spaces and initial configurations, how predictable and efficient is that dynamics? It turns out that some aspects of MAJ CA dynamics are relatively complex; for example, for many cellular spaces and rule radii r, even in the simplest 1-dimensional case, MAJ CA have *exponentially many* fixed-point (FP) configurations (Tosic and Agha 2004a). However, it turns out that convergence to each of those FPs happens rather quickly, in time that is linear in the number of CA's nodes, that is, O(n). Moreover, this fast convergence takes place from all initial configurations (and, as it turns out, is even faster if the nodes update sequentially one at a time, than if they update synchronously in parallel (Tosic and Agha 2004a, b; Tosic 2010a)). Therefore, the dynamics of these MAJ CA, while nontrivial, is always very efficient. Moreover, this efficiency holds for the higher dimensional underlying cellular spaces, as well.

Insofar as the MAJ CA's predictability, the guaranteed fast convergence to a FP regardless of the starting configuration (as well as the ability to characterize analytically, under what circumstances and from which initial configurations a temporal cycle might be reachable) indicates that these networks are also inherently *globally predictable*, as well. In particular, given an arbitrary starting configuration and the cellular space ("network topology"), not only can we effectively determine whether the system will evolve to a FP (and how quickly!), but we can also tell to which among in general exponentially many FPs will the system converge (Tosic and Agha 2004a).

While the step-by-step simulation is quite efficient in this particular case, it turns out that these questions about the "ultimate destiny" of the system evolving from a given starting state can also be answered analytically; see our prior work (Tosic and Agha 2004a; Tosic 2010a; Tosic and Agha 2004b) for details. We remark that this $O(\text{n})$ worst-case rate of convergence by MAJ CA to some FP has been established analytically. For some simulation-based results on what this linear time (indeed, in many cases, sublinear time) converge to a stable configuration typically looks like in practice, see Tosic and Raju (2011).

The AND, OR, and MAJ local update rules all belong to a class of arguably the simplest local update functions, called *simple threshold rules* (Tosic and Agha 2004a). In the 1-D CA cases, each simple threshold update rule can be defined as *"at-least-l-out-of-k"* for appropriate values of nonnegative integers l and k. That is, a node will update its state to 1 if and only if at least l out of its k input values are 1; and it will update to 0, otherwise. So, for example, for a CA with rule radius r, the MAJ update rule can be described as "at-least-(r+1)-out-of-(2r+1)." It turns out that all these CA with simple threshold (Boolean-valued) functions for their local update rule have asymptotic dynamics that are both predictable and efficient in our sense (Tosic and Agha 2004a; Tosic 2006a, 2010a). In fact, most of these update rules lead to a small number of different possible dynamics; the only exception is the MAJ rule which, as discussed earlier, results (in general) in a number of fixed points, and therefore a total number of different possible evolutions, that is, exponential in the number of CA's nodes n (Tosic and Agha 2004a; Tosic 2010a).

Here is what has been known about the dynamics of simple threshold CA, summarizing the most relevant for our present purposes results from Tosic and Agha (2004a), Tosic (2010a), Tosic (2017), and Tosic and Agha (2004b):

Theorem 2 For all simple threshold CA, the following properties of their asymptotic dynamics hold:

- The reachability and FP reachability questions, for any starting configuration, can be answered efficiently, via step-by-step simulation. In particular, all these CA are both efficient and locally predictable.
- For all simple threshold rules (that is, update rules of the form *at-least-l-out-of-k*), except for the *majority* rule, there is a small number of FPs, and those FPs can be determined efficiently—either analytically or via low-degree polynomial, typically $O(\text{n})$, step-by-step simulation.
- For any configuration C, it can be efficiently determined to which FP's *basin of attraction* C belongs. In particular, all these simple threshold CA (with a possible exception of MAJ CA) also have globally predictable dynamics.

- The MAJ CA in general have a number of FPs that are exponential in the CA's number of nodes n. However, given the cellular space and rule radius r, all these FPs can be analytically determined.
- Given an arbitrary configuration of a MAJ CA, it can be effectively determined to which fixed point's *basin of attraction* it belongs, and the transient chain leading to that FP is of length at most $O(n)$.
- In particular, the MAJ CA are efficient and globally predictable, just like the rest of the simple threshold CA (although these CA with the *majority* rule do have exponentially many more FPs, and therefore exponentially more possible evolutions, than the CA on the same cellular spaces using any other simple threshold rule $f \neq MAJ$).

How complex can the Boolean network dynamics get, once we generalize from the classical CA to more general BN models, for the same restricted class of local interactions as encapsulated by these Boolean-valued simple threshold update rules? If we restrict the local update rules to simple threshold functions, and we require that all the nodes use the same update rule, then making the network topology more complex than the most common, really simple cellular spaces such as 1-D lines or rings, 2-D rectangular grids, or tori, actually will not make much of a difference—the resulting BN dynamics will still remain, in general, both predictable and efficient.

However, once more complex network topologies are combined with not all the BN's nodes having to update their respective states according to the same local rule, the dynamics complexity threshold gets quickly crossed. In particular, *once the underlying BN's nodes are allowed to use just two different update rules, even if both rules are still restricted to simple threshold functions, the resulting dynamics can become quite complex* (Tosic 2010b, 2016, 2017). This includes the problem of determining the total number of possible evolutions (i.e., different asymptotic behaviors from all possible initial configurations) becoming computationally intractable—formally, **#P**-complete.

Moreover, it turns out that such complex behaviors can be obtained even on uniformly sparse graphs, as established by Tosic (2010b) and Tosic (2017) for the BN models using an agent-based modeling and as an abstraction of large-scale multi-agent systems, and by Tosic (2016), in the context of discrete Hopfield networks as a model for the storage capacity of an associative memory. These remarks, based mostly on our own prior work on BNs only "slightly more general" than the classical finite CA, are a good transition point for us to wrap up the discussion of those Boolean networks that are both efficient and predictable, and briefly overview several classes of similar networks that are provably unpredictable, some of which as it turns out also inefficient in our sense, whereas others, while still unpredictable, actually turn out to be quite efficient.

4 Boolean Networks with Globally Unpredictable Dynamics

That the dynamics of a sufficiently general BN will in general be both unpredictable and inefficient—or at least unpredictable and inefficient starting from some of the initial configurations—is not surprising. By "sufficiently general," we mean with respect to any or all of the three key parameters characterizing the structural complexity of a discrete network: i) the *underlying graph* (called *cellular space* in case of CA, and referred to as *network topology* for the more general BNs); ii) the *local update rules* f_i; and iii) the *degree of diversity of local interactions* among the agents, that is, how many different local update rules are different nodes of a BN allowed to choose from; to differentiate these BN and DHN models from the classical CA, we assume that there are at least two different local update rules used by the nodes in the underlying graph of a BN or a DHN.

It is well known that allowing some complexity with respect to ii) alone can lead to rather complex—and, in particular, unpredictable and inefficient in our sense—long-term dynamics, as shown for the classical one-dimensional CA with a sufficiently general local update rule (Wolfram 1984, 1986). However, as discussed in the previous section, once the classical CA have appropriately restricted local update rules, such as requiring that the nodes update according to a (one and the same across all the nodes) simple threshold function or a monotone Boolean-valued function, then the resulting dynamics of such restricted classes of CA generally become both efficient and globally predictable.

On the other hand, generalizing rather minimally with respect to (i) and (iii), while imposing restrictions on the complexity of local update rules in (ii), has more recently also been shown to potentially lead to rather complex dynamics. A *minimal generalization* with respect to criterion (iii) simply means allowing each network node to choose one of the two different update rules, rather than every node having to update according to the same update rule (Tosic 2010b). Criterion (i) about the underlying graphs, on the other hand, is more complex: these underlying graphs of dynamical networks can be generalized, starting with regular Cayley graphs of classical CA, in many different ways. The "minimal generalization" we have studied in our research focuses on allowing more general underlying graphs that are however still required to be *uniformly sparse*, that is, where each node has $O(1)$ neighbors. In fact, we have succeeded in obtaining highly complex network dynamics, even when the underlying BN's graph is required to be not just uniformly sparse, but also regular—for example, such that the number of neighbors is deg(v) = 3 for each node v in the network (Tosic 2010b, 2017). Such sparseness can also be imposed in the context of DHNs, by requiring that all the weights w_{ij} except for a few (say, only 3 or 4 per each row of the weight-matrix W) be set equal to 0. Another line of modest generalization that we have investigated is to consider graphs where exactly one node is allowed to have a large neighborhood—that is, BNs and DHNs defined over wheel-like and starlike graphs (Tosic 2006b, 2010b).

Limiting our focus to the network dynamics efficiency and predictability, we summarize several results about the computational (in)tractability of the reachability-flavored problems, as well as of enumeration of possible asymptotic dynamics (as a function of the choice of the initial configuration) problems. We recall that, in systems that have mostly (or exclusively) fixed points and no temporal cycles, the FP reachability question as defined in Sect. 2 becomes the following: starting from an arbitrary initial configuration, can we tell to which (among possibly many) FPs will the system's dynamics eventually converge? Examples of systems with no temporal cycles include sequential CA with simple threshold update rules (Tosic and Agha 2004a, b), as well as certain types of discrete Hopfield networks (Tosic 2016).

So, what are some results on the FP and other types of reachability, for appropriately restricted subclasses of BNs and DHNs, which would immediately imply both unpredictability and inefficiency of those systems' asymptotic dynamics?

Theorem 3 (cf. based on (Barrett et al. 2003)): The reachability problem for an arbitrary BN whose nodes are restricted to update according to the Boolean-valued threshold update rules (and different nodes are in general allowed to use different such rules) is in the worst-case **PSPACE**-complete.

We note that the *linear threshold functions* in the work of Barrett et al. (2003) are defined as is common in, say, most of the Hopfield network literature; however, they are more general than the simple threshold functions discussed in the context of *majority* and related CA models in the previous section. In particular, the (not necessarily simple) linear threshold functions in general may have different weights associated with different nodes; see Definition 2 and reference Tosic (2016).

This result immediately implies both general unpredictability and inefficiency of the dynamics of such BNs. To see why such systems are also inefficient in our sense, for at least some initial configurations, we use a simple argument by contradiction: if these systems were actually always efficient (starting from an arbitrary initial state), then such a system would converge to its ultimate stationary behavior relatively quickly, and certainly in a number of steps at most polynomial in n, contradicting the **PSPACE**-completeness of determining what will that asymptotic behavior be.

For sufficiently general deterministic discrete systems, therefore, it is computationally intractable to determine, starting from an arbitrary initial state, whether a FP or a temporal cycle will be reached. Many systems, however, have few or no temporal cycles—but may have many FPs. To argue that such systems are predictable, it does not suffice to say that we know the system will converge to an FP—we also want to be able to tell to which (among possibly exponentially many) FP will it converge, given the initial configuration. That is, we want an answer to the more refined and generally harder FP-reachability (or, where applicable, CC-reachability) problem(s). The closely related computational problems to the FP reachability (and CC reachability) are those of i) determining how many FPs a given system has; ii) determining the size of each FP's basin of attraction; and iii) given an arbitrary configuration, to which FP's basin of attraction does that configuration belong.

It turns out that answering these questions becomes computationally intractable for even much more restricted Boolean networks than those in Theorem 3. Concretely:

Theorem 4 (cf. based on (Tosic 2006b)) The following problems are all **#P**-complete, even when the underlying graphs of a BN or DHN are restricted to star graphs or wheel graphs, and all local update rules are monotone Boolean functions:

1. Determining the exact number of all FPs
2. Determining the exact number of all TCs
3. Determining the exact number of Gardens-of-Eden
4. Determining the total number of predecessors of an arbitrary configuration
5. Determining the total number of all ancestors of a given configuration, and, in particular, determining the size of the "basin of attraction" of a given FP

Note that the established computational complexity of the above properties, as summarized in Theorem 4, still does not imply that those systems are necessarily *inefficient.* Namely, it is possible that, for an arbitrary starting configuration, the convergence to the eventual stationary pattern actually takes place fast, implying the underlying dynamics' efficiency. That convergence is typically to some FP configuration since, depending on the exact details of a particular BN including its communication model, the temporal cycles either do not exist at all or else are very few and far between (as proved and discussed in some detail by Tosic (2010a) and Tosic (2011)).

Needless to say, for sufficiently general BNs, specific instances and even entire classes can be found where this convergence is not fast, implying inefficiency in general. However, with some further restrictions on the update rules, such as that they all be monotone or of a simple threshold variety, the convergence to a FP can be shown to actually always be fast, starting from any initial configuration—resulting in systems with potentially many possible dynamics, yet all of whose dynamics are actually efficient. As we have seen with MAJ CA, this exponential number of possible evolutions need not by itself imply high complexity or (global) unpredictability. However, for a number of interesting classes of BNs and DHNs, determining the actual number of ways in which the system may evolve, as a function of the starting configuration and the exact details of the specific update rule at each node and the structure of the underlying graph, is in the worst case provably computationally intractable. Therefore, those systems' asymptotic dynamics are generally globally unpredictable in the worst case—yet, due to fast convergence to an FP from each starting configuration, the said dynamics are still efficient. A specific example of a class of unpredictable yet efficient systems is found among certain discrete Hopfield networks; see Tosic (2016).

Our last theoretical results, establishing complex dynamics of descriptively rather simple and structurally very constrained systems, and in particular showing global unpredictability of their dynamics in general, with the question of efficiency being still open, are summarized in the following theorem (cf. based on our prior work and key theoretical results originally established in Tosic (2010b) and Tosic (2017)):

Theorem 5 Let a *Boolean network* (BN) be defined over a uniformly sparse graph, and let all BN's local update rules be simple threshold functions. Enumerating all possible dynamics of such a BN is **#P**-complete (and, therefore, under the usual assumptions in the theory of computing, *computationally intractable*), even when these two restrictions simultaneously hold:

i. Each node in the underlying graph has only three neighbors.
ii. Each node "gets to choose" from just two different simple threshold rules.

We again point out that, while computational hardness of the reachability-flavored problems in general does imply system's global unpredictability, the converse, that is, computational tractability of the reachability problems (incl. FP and CC reachability), need not necessarily imply global predictability, since not only there could be exponentially many different possible dynamics (across all possible starting configurations), but also, as Theorem 5 establishes, it may be computationally hard to determine in how many different ways the system may evolve, and how many different stationary patterns it may have.

Further, complexity of the enumeration problems pertaining to FPs, CCs, and also sizes of basins of attractions, in general, also leaves the efficiency problem still open: starting from an arbitrary initial state, the convergence to a stationary pattern, whatever that pattern may be, could be, but does not have to be, fast (where fast convergence implies a short transient chain and therefore efficiency). In our future work, we will investigate more thoroughly which among the globally unpredictable systems lead to efficient vs. which ones lead to inefficient dynamics.

5 Summary and Conclusions

We have been modeling the emerging behaviors and collective dynamics of large-scale multi-agent and cyber-physical systems as networks of deterministic communicating finite-state machines—in particular, using the classical (finite) cellular automata and their generalizations such as Boolean networks and discrete Hopfield networks (Tosic and Agha 2005; Tosic 2005, 2011, 2017, 2002; Tosic and Ordonez 2017). In this chapter, our focus is on *predictability* and *efficiency* of those networks' dynamics. We define predictability and efficiency of such systems' deterministic dynamics in strictly computational terms, that is, with respect to the required computational costs of answering key questions about those systems' *asymptotic behavior*. We then discuss which classes of deterministic discrete networks are (un)predictable and which ones are (in)efficient with respect to our definitions.

In that context, we have reinterpreted a number of mostly theoretical (and a few simulation-based) results about the asymptotic dynamics of certain classes of

CA and BNs, in terms of the proposed notions of efficiency and predictability. We discuss several classes of such discrete deterministic networks whose dynamics fall into different categories with respect to their efficiency and predictability, ranging from those that are both efficient and predictable to those that are neither, with a few classes of systems whose dynamics lie "somewhere in between." In particular, we observe that there are interesting classes of CA and BNs in each of the four "quadrants" with respect to the two most fundamental properties of interest, namely, efficiency and global predictability. In this chapter, we have mostly focused on some examples of systems in the first and fourth quadrants, and started discussing globally unpredictable systems whose efficiency ranges from efficient to provably (under the usual assumptions in theoretical computer science) inefficient to those with "efficiency status unknown."

In future work, we plan to systematize the existing knowledge about various discrete networks' asymptotic dynamics in terms of their efficiency and predictability, as well as determine the status of (in)efficiency for several classes of Boolean network models known to be globally unpredictable, yet for whose dynamics it is, to the best of our knowledge, still unknown whether or not they are efficient in our sense.

Acknowledgments The author would like to thank Georgi Georgiev and Kevin Vixie for their feedback, patience, and support.

References

Bandini S, Manzoni S, Simone C (2002) Heterogeneous agents situated in heterogeneous spaces. Appl Artif Intell Int'l J 16(9–10):831–852

Barrett C et al (2001) Gardens of Eden and fixed points in sequential dynamical systems. Discrete Math Theor Comp Sci AA:95–110

Barrett C et al (2003) Reachability problems for sequential dynamical systems with threshold functions. Theor Comput Sci 295(1–3):41–64

Davey N, Calcraft L, Adams R (2006) High capacity, small world associative memory models. Connect Sci 18(3):247–264

Floreen P, Orponen P (1989) On the computational complexity of analyzing Hopfield nets. Complex Syst 3:577–587

Graudenzi A et al (2011) Dynamical properties of a Boolean model of a gene regulatory network with memory. J Comp Biol 18(10)

Hopfield J (1982) Neural networks and physical systems with emergent\collective computational abilities. Proc Natl Acad Sci 79:2554–2558

Hopfield J, Tank D (1985) Neural computation of decisions in optimization problems. Biol Cybern 52:141–152

Kauffman SA (1984) Emergent properties in random complex automata. Phys D Nonlinear Phenomena 10(1–2):145–156

Mortveit H, Reidys C (2001) Discrete sequential dynamical systems. Discret Math 226(1–3):281–295

Orponen P (1996) Computational complexity of neural networks: a survey. Nordic Journal of Computing

Sima J, Orponen P (2003) General-purpose computation with neural networks: a survey of complexity theoretic results. Neural Comput 15(12):2727–2778

Tosic P (2002) Some examples of NP-complete and CoNP-complete problems about certain configuration space properties of SDSs and SyDSs. IL, USA, Urbana. (unpublished)

Tosic P (2005) Cellular automata for distributed computing: models of agent interaction and their implications. In: Proc. IEEE Int'l conf. systems, man & cybernetics (SMC'05), pp 3204–3209

Tosic P (2006a) On modeling and analyzing sparsely networked large-scale multi-agent systems with cellular and graph automata. In: Alexandrov V et al (eds) Comput. science ICCS-2006: 6th Int'l conf. (proc. part III), LNCS, vol 3993. Springer, pp 272–280

Tosic P (2006b) On the complexity of counting fixed points and gardens of Eden in sequential & synchronous dynamical systems. Int'l J Found Comput Sci 17(5):1179–1203

Tosic P (2010a) Cellular automata communication models: comparative analysis of parallel, sequential and asynchronous CA with simple threshold update rules. Int'l J Nat Comput Res 1(3):66–84

Tosic P (2010b) On the complexity of enumerating possible dynamics of sparsely connected Boolean network automata with simple update rules. In: Discrete mathematics and theoretical computer science (DMTCS), pp 125–144

Tosic P (2011) Modeling large-scale multi-agent systems with sequential and genuinely asynchronous cellular automata. Acta Phys Polonica B (Proc Suppl) 4(2):217–236

Tosic P (2016) On simple models of associative memory: network density is not required for provably complex behavior. In: Proc. brain informatics and health, lecture notes in computer science (LNCS) series, vol 9919. Springer, pp 61–71

Tosic P (2017) On phase transitions in dynamics of cellular and graph automata models of sparsely interconnected multi-agent systems. In: ACM proc. autonomous agents & multi-agent systems (AAMAS'17), Sao Paulo, Brazil, pp 474–483

Tosic P, Agha G (2004a) Characterizing configuration spaces of simple threshold cellular automata. In: Proc. 6th Int'l Conf. On cellular automata for Research & Industry (ACRI'04). Springer, Amsterdam, The Netherlands, pp 861–870

Tosic PT, Agha G (2004b) Concurrency vs. sequential Interleavings in 1-D threshold cellular automata. In: Proc. of the 18th IEEE Int'l parallel and distributed processing symposium (IPDPS'04) workshop on advances in parallel and distributed computing models (APDCM), Santa Fe, New Mexico, USA

Tosic P, Agha G (2005) On computational complexity of counting fixed points in symmetric Boolean graph automata. In: Proc. 4th Int'l con. unconventional computing (UC'05), LNCS, vol 3699. Springer, Seville, Spain, pp 191–205

Tosic P, Ordonez C (2017) Boolean network models of collective dynamics of open and closed large-scale multi-agent systems. In: Proc. 8th Int'l conf. industrial applications of Holonic & Multi-Agent Systems (HoloMAS'17), Lyon, France, pp 95–100

Tosic PT, Raju SNV (2011) On convergence properties of one-dimensional cellular automata with majority cell update rule. In: Proc. Int'l conference on scientific computing (CSC-11), Las Vegas, Nevada

Wolfram S (1984) Universality and complexity in cellular automata. Phys D Nonlinear Phenomena 10(1–2):1–35

Wolfram S (1986) Theory and applications of cellular automata. World Scientific, Singapore

Wolfram S (1994) Cellular automata and complexity (collected papers). Addison-Wesley, Reading, UK

Efficiency of Grammars for Natural Languages

Relja Vulanović

Abstract A general approach to measuring grammar efficiency of natural languages is described and applied to a model of the Turkish part-of-speech system. It is shown that an earlier grammar-efficiency formula may be relatively complicated to use. A simplification is introduced which uses a recently proposed measure of the degree of violation of the one-meaning–one-form principle.

Keywords One-meaning-one-form principle · Bijection · Relation · Formal grammar · Grammar efficiency · Propositional function · Parts-of-speech system · Turkish

1 Introduction

Any system which converts some input into an output is more efficient if it needs a smaller amount of input to create a greater amount of useful output. Therefore, the efficiency of such a system can be defined like machine efficiency: it should be directly proportional to a measure of the useful output and inversely proportional to a measure of the total input. I applied this approach to linguistics for the first time in Vulanović (1991), where I defined and measured grammar efficiency of natural languages. Linguistic input consists of all grammatical forms and categories, called *grammatical conveyors*, that the language uses to convey any information which is deemed important in the language (*linguistically relevant information*). This information constitutes linguistic output. Grammatical conveyors include word classes (parts of speech), cases and other nominal forms, verbal forms, etc. Let C denote the set[1] of grammatical conveyors. Linguistically relevant information is

[1] All sets in this paper are finite and non-empty, unless stated otherwise.

R. Vulanović (✉)
Department of Mathematical Sciences, Kent State University at Stark, North Canton, OH, USA
e-mail: rvulanov@kent.edu

G. Y. Georgiev, M. Shokrollahi-Far (eds.), *Efficiency in Complex Systems*,
Springer Proceedings in Complexity, https://doi.org/10.1007/978-3-030-69288-9_7

formally described as a set I, containing elements like syntactic functions (subject, object, predicate, etc.), semantic functions (beneficiary, goal, etc.), and propositional functions (heads and modifiers of propositional and referential phrases).

Let Eff denote a numerical quantity which measures grammar efficiency. Then, according to the above discussion,

$$\text{Eff} \propto \frac{|I|}{|C|},$$

where the number of elements in a set A is denoted by $|A|$.

Linguistically relevant information can also be conveyed by word order.[2] Word order belongs to syntax. Syntactic rules tell us how to combine grammatical conveyors to produce sentences which carry the intended information. Therefore, word-order rules cannot be represented as elements of the set C. However, a grammar is more efficient if it has less rules, including the word-order rules. This is represented in the formula for Eff by the coefficient K,

$$\text{Eff} = \kappa K \frac{|I|}{|C|}, \tag{1}$$

where κ is another coefficient, the purpose of which is just to scale the formula. The coefficient K is inversely proportional to the number of word-order and other rules which convert the input to the output. It will be shown that the input-to-output conversion mechanism is the most important component of grammar efficiency.

The formula in (1) is a general formula of the type used in Vulanović (2003, 2007). These two papers modify the formula from Vulanović (1993), which is itself a modification of the original formula in Vulanović (1991). Yet another modification is introduced in the present paper with the aim of simplifying the calculations needed to evaluate grammar efficiency. This is done by incorporating in the coefficient K a recent result from Vulanović and Ruff (2018), where it is proposed how to measure the degree of violation of the one-meaning–one-form principle, Anttila (1972).

The latest efficiency formula, that from Vulanović (2007), is presented in Sect. 3 and illustrated by applying it to a model of the Turkish parts-of-speech (PoS) system. A formal description of PoS systems is given before that, in Sect. 2. The Turkish PoS system provides a motivation for a simplification of the formula. The simplified formula is introduced in Sect. 4 and applied to the same PoS system. Section 5 offers some concluding remarks.

[2]For instance, consider the English sentence *Brutus killed Caesar*. When the words *Brutus* and *Caesar* are interchanged, the meaning of the sentence is changed.

Table 1 Some word classes and the propositional functions they fulfill

Word Class	P	R	r	p
V = verbs	+			
N = nouns		+		
a = adjectives			+	
m = manner adverbs				+
H = heads	+	+		
M = modifiers			+	+
Λ = non-verbs		+	+	+
C = contentives	+	+	+	+

2 Part-of-Speech Systems

Hengeveld's approach to PoS systems is followed here; see Hengeveld (1992) and also Hengeveld et al. (2004), Hengeveld and van Lier (2010). Four propositional functions (syntactic slots) are considered in this approach:

- P = head of predicate phrase,
- p = modifier of predicate phrase,
- R = head of referential (nominal) phrase,
- r = modifier of referential phrase.

Word classes are distinguished by the propositional functions they can fulfill. Table 1 shows some word classes. Verbs, nouns, adjectives, and manner adverbs—each of these word classes is called *rigid* because each has exactly one propositional function. However, there are many languages that have *flexible* word classes, those that have more than one propositional function. Heads, modifiers, non-verbs, and contentives are examples of flexible word classes in Table 1.

A PoS system consists of several word classes that are used to covey the propositional functions. One or both head modifiers, p or r, may be missing, but only systems with all four propositional functions are considered here. Turkish has such a PoS system and it uses V, Λ, and M for its word classes, Hengeveld and van Lier (2010). The Turkish PoS system is analyzed in the next two sections.

3 Grammar Efficiency: The Old Formula

How efficient is the Turkish PoS system? The question is answered in this section using the efficiency formula from Vulanović (2007). The formula is like in (1) with the coefficient K equal to the so-called *parsing ratio* Q. Before Q is defined below, the sets I and C can be identified in the Turkish PoS system as

$$C = \{\mathrm{V}, \Lambda, \mathrm{M}\} \quad \text{and} \quad I = \{\mathrm{P}, \mathrm{R}, \mathrm{r}, \mathrm{p}\}.$$

Therefore, $|C| = 3$ and $|I| = 4$.

In the model of the Turkish PoS system, which is considered here, only simple intransitive sentences are represented. Such simplifications are necessary in any linguistic model. According to Miestamo (2008), in the absence of complete detailed grammars of entire languages, we must restrict our studies to the *local* grammar complexity, or efficiency. The word *local* refers here to specific separate aspects of grammar across languages. Thus, we can talk about complexity/efficiency of the morphological system, of simple sentences, of negation, etc.

A sentence is represented simply as a string of grammatical conveyors, that is, word-class symbols. Not every such string is a sentence. A sentence has to convey the information which identifies the heads P and R and, optionally, the modifiers p and r. For simplicity, only continuous predicate and referential phrases are considered here, so that the modifier, if it is present in the sentence, has to stand next to its head. Therefore, a sentence has to be interpreted as at least one of the 18 strings of propositional functions listed below:

$$\left.\begin{array}{l}\text{PR or RP,}\\ \text{PRr, PrR, RrP, or rRP,}\\ \text{PpR, pPR, RPp, or RpP,}\\ \text{PpRr, PprR, pPRr, pPrR, RrPp, RrpP, rRPp, or rRpP.}\end{array}\right\} \tag{2}$$

The strings which are actually permitted in the language form a set which is denoted by R. In the present model of the Turkish PoS system, this set is

$$R = \{\text{RP}, \text{rRP}, \text{RpP}, \text{rRpP}\}, \tag{3}$$

which indicates that reference comes before predication and modifiers before heads. This corresponds to the basic word order in Turkish, Hengeveld et al. (2004).

A string of grammatical conveyors is interpreted (analyzed, parsed) based on all possible propositional functions each conveyor can fulfill. The propositional functions of V, Λ, and M, as the word classes in the Turkish PoS system, are given in Table 1. This can also be represented by a relation Φ, $\Phi \subseteq C \times I$, which in this case looks as follows:

$$\Phi = \{(\text{V}, \text{P}), (\Lambda, \text{R}), (\Lambda, \text{r}), (\Lambda, \text{p}), (\text{M}, \text{r}), (\text{M}, \text{p})\}. \tag{4}$$

It is also convenient to write Φ as a mapping (which in Turkish is not a function) from C onto I,

$$\Phi : \ \text{V} \to \text{P}, \ \ \Lambda \to \text{R}, \text{r}, \text{p}, \ \ \text{M} \to \text{r}, \text{p}. \tag{5}$$

In general, a grammar G is determined by the sets C and I, the relation Φ, and the set R: $G = (C, I, \Phi, R)$. In the case of PoS systems, R has to contain at least one string from each of the four lines in (2). The efficiency of a particular grammar G is written as $\text{Eff}(G)$.

Having (4) or (5) in mind, we can see that the string ΛMV is a sentence in the model of the Turkish PoS system because it can be interpreted as RpP. Moreover, this is an unambiguous sentence since RpP is its only interpretation. Formally, the analysis of the sentences is carried out[3] from left to right, one symbol at a time. The length of the input string is not known in advance and, as for the order of the propositional functions, it is only used that the modifiers have to appear next to their heads. Thus,

$$\Lambda\text{MV} \rightarrow \text{RrP} \mid \underline{\text{RpP}} \mid \text{r}- \;\mid \text{p}-, \tag{6}$$

where all possible parses are attempted and only the first two are completed, a hyphen indicating that the parse is abandoned.[4] Initially, there are two possible interpretations of the sentence ΛMV, but, when parsing is done, the completed parses are compared to the strings in the set R in (3). The string RrP is not in R and is discarded. At the same time, RpP, being in R, is kept and, being the only interpretation, it is underlined in (6). Had RrP been in R, the sentence ΛMV would have been ambiguous, having two interpretations. This kind of use of the set R, after all possible parses are done, is similar to the *regulated rewriting* of Dassow and Păun (1989). The reason for this setup is to represent how complex the relation Φ is. The further away it is from a bijection between C and I, the more the one-meaning–one-form principle is violated and the greater the number of parsing attempts.

There are eight sentences in the Turkish Pos system:

$$\Lambda\text{V}, \Lambda\Lambda\text{V}, \text{M}\Lambda\text{V}, \Lambda\text{MV}, \Lambda\Lambda\Lambda\text{V}, \text{M}\Lambda\Lambda\text{V}, \Lambda\Lambda\text{MV}, \text{M}\Lambda\text{MV}. \tag{7}$$

Among them, there is one ambiguous sentence, ΛΛV:

$$\Lambda\Lambda\text{V} \rightarrow \text{RrP} \mid \underline{\text{RpP}} \mid \underline{\text{rRP}} \mid \text{p}- . \tag{8}$$

Let ρ indicate the number of sentences and let ρ_0 be the number of ambiguous sentences. In the Turkish PoS system, $\rho = 8$ and $\rho_0 = 1$. One more count is needed to define the parsing ratio. It is the number, denoted by ρ', of all parsing attempts of all sentences that can be parsed like any of the strings in (2). Note that the count ρ' also includes sentences which may not be permitted by the set R. The reason for this is to indicate how restrictive the set R is in comparison to all possible strings that could be permitted.

The parsing ratio is defined as

$$Q = \frac{\rho - \rho_0}{\rho'}, \tag{9}$$

[3]There is no intention here to simulate how the human parser works.

[4]The last two parses are abandoned because, after interpreting the leading Λ as a modifier, it is expected that the next propositional function be the corresponding head, but M cannot function as a head.

giving

$$\text{Eff} = \kappa Q \frac{|I|}{|C|} = \kappa \frac{\rho - \rho_0}{\rho'} \frac{|I|}{|C|} \tag{10}$$

for the grammar-efficiency formula. The formula for Q is motivated by the following considerations. A grammar is more efficient

(a) if it permits more sentences, i.e., if ρ is greater (the number of word-order rules is smaller),
(b) if the number of sentences ρ is closer to ρ', i.e., ρ/ρ' is greater (word order is more flexible),
(c) if the possible sentences (before the word-order-related rules in R are applied) are easier to parse, i.e., if Φ is closer to a bijection, which means that ρ' is less, and
(d) if it has fewer ambiguous sentences, i.e., if ρ_0 is less.

Whereas the counts ρ and ρ_0 are easy to obtain for the Turkish PoS system, calculating ρ' is more cumbersome. It turns out that 32 sentences need to be parsed and their parsing attempts counted together. For instance, one of those sentences is ΛVΛM and its parsing attempts are

$$\Lambda\text{V}\Lambda\text{M} \to \text{RPp} - \mid \text{r} - \mid \text{pPRr} \mid \text{pPr} - .$$

This sentence is not interpreted as any of the strings in the set R, but its four parsing attempts have to be included in the count ρ'. The final result is $\rho' = 100$, giving

$$Q = \frac{7}{100} \tag{11}$$

for the parsing ratio in Turkish.

It remains to explain how the scaling coefficient κ is determined in (10). The formula is scaled so that Eff $= 1$ for maximally efficient grammars in the family $\Gamma = \Gamma(|C|, |I|)$ of grammars which have the same number of grammatical conveyors, $|C|$, to convey the same amount of linguistically relevant information, $|I|$. The grammar G^* is called the *maximally efficient grammar* in Γ if

(i) its relation Φ uses every element of I exactly once,
(ii) it does not permit any ambiguous sentence, and
(iii) it has the greatest value of the coefficient K.

Suppose G^* exists[5] and let Q^* be the value of its parsing ratio, so that Q^* is the greatest value of K. Then, define

$$\kappa = \frac{|C|}{|I| Q^*} \text{ so that Eff}(G^*) = 1.$$

[5]This is not always the case. If $|C| > |I|$, (i) cannot be achieved. Also, if, for instance, $C = \{\text{C}\}$, (ii) is impossible even if the set R is maximally restricted. However, if there is a grammar in Γ satisfying the conditions (i) and (ii), then G^* exists.

With this choice of κ, it follows from (10) that

$$\mathrm{Eff}(G) = \frac{Q}{Q^*} \tag{12}$$

for any grammar $G \in \Gamma$. On the other hand, if G^* does not exist, set $\kappa = 1$ (the largest possible value of κ in the above case) so that

$$\mathrm{Eff}(G) = Q\frac{|I|}{|C|}.$$

The formula (12) shows why the coefficient K (here, $K = Q$) is the most important component of the efficiency formula (1). At the same time, (12) indicates that the measure of grammar efficiency is a relative measure in the sense that what is achieved in the grammar is compared to the maximum potential. In this way, the use of the scaling coefficient κ enables a more realistic comparison of efficiencies of grammars with different amount of conveyors and information conveyed. The scaling coefficient has the effect of the same yardstick used to measure grammar efficiency of grammars which may be very different.

Finding κ for the Turkish PoS system is also relatively complicated. All grammars in the family Γ with $|C| = 3$ and $I = \{\mathrm{P}, \mathrm{R}, \mathrm{r}, \mathrm{p}\}$ should be explored and the maximally efficient one should be found. This has already been done in Vulanović (2008) and the result is that such a grammar exists; it is the grammar with $C = \{\mathrm{H}, \mathrm{a}, \mathrm{m}\}$, for which $Q^* = \frac{5}{8}$. Therefore, it follows from (11) and (12) that the efficiency of the Turkish PoS system is

$$\mathrm{Eff} = \frac{7}{100} \cdot \frac{8}{5} = \frac{14}{125} = 0.112.$$

This is relatively low because of the overlapping roles of Λ and M and because of the fixed word order.

Since the procedure for evaluating ρ' and κ (or Q^*) may be very involved, it is of interest to find another way of calculating the coefficient K in the formula (1).

4 Grammar Efficiency: A New Formula

Items (b) and (c) in the above list indicate the dual role of the quantity ρ' in the parsing ratio Q, as defined in (9). On the one hand, ρ' is part of the measure of word-order flexibility/rigidity and, on the other hand, it shows how close or far the relation Φ is from a bijection. Since the latter is in fact unrelated to parsing, it is not natural to use a parsing-based count like ρ' to measure this feature. A relatively simple, parsing-independent, formula for measuring how far a relation is from a bijection has been recently proposed in Vulanović and Ruff (2018). It is shown in this section how to incorporate this formula into a measure of grammar efficiency and simplify the procedure for calculating the parsing ratio Q.

Let B denote a subset of Φ containing all one-to-one ordered pairs. The Turkish model can be used to illustrate what is meant by this; in Turkish, $B = \{(\mathrm{V}, \mathrm{P})\}$. The measure of how far Φ is from a bijection is denoted by $\mu(\Phi)$ and is defined as

$$\mu(\Phi) = \frac{2|\Phi| - |B|}{\min\{|C|, |I|\}}. \tag{13}$$

The formula (13) is motivated by the consideration that Φ is further away from a bijection if $|\Phi|$ is greater while $|C|$ and $|I|$ are smaller. Also, if Φ is a bijection, then

$$|\Phi| = |B| = |C| = |I| \quad \text{and} \quad \mu(\Phi) = 1.$$

Otherwise, $|\Phi|$ is strictly greater than $|B|$ and at least one of $|C|$ or $|I|$, and then (13) implies that

$$\mu(\Phi) > \frac{|\Phi|}{\min\{|C|, |I|\}} > 1.$$

In the Turkish PoS system, $|\Phi| = 6$, $|B| = 1$, $|C| = 3$, and $|I| = 4$, thus

$$\mu(\Phi) = \frac{2 \cdot 6 - 1}{\min\{3, 4\}} = \frac{11}{3}. \tag{14}$$

If the relation Φ is a bijection, each sentence has exactly one possible interpretation and the total number of parsing attempts is simply equal to the number of sentences. In this situation, the grammar may permit all permutations of all possible sentences, but it also may restrict word order via the strings in the set R. The flexibility of word order alone can be measured by a quantity simpler than the parsing ratio Q. Let this simpler measure, denoted by $\bar{Q}$, be defined by

$$\bar{Q} = \frac{\rho - \rho_0}{\bar{\rho}}, \tag{15}$$

where $\bar{\rho}$ is the number of all possible unambiguous sentences. Thus, this formula represents the number of sentences permitted by the grammar, relative to the maximum number of sentences that could be permitted theoretically. In the Turkish PoS system, the 32 possible sentences should still be considered and parsed to determine how many of them are ambiguous, but it is not necessary to find and count every parsing attempt like before. It turns out that 9 of the 32 sentences are ambiguous and, therefore, $\bar{\rho} = 23$, giving

$$\bar{Q} = \frac{7}{23}. \tag{16}$$

The grammar is more efficient if $\bar{Q}$ is greater and if $\mu(\Phi)$ is less. Therefore, the coefficient K in (1) can be defined as

$$K = q := \frac{\bar{Q}}{\mu(\Phi)}. \tag{17}$$

The scaling coefficient κ can be found analogously to the procedure describe in Sect. 3. The grammar G^* is called the maximally efficient grammar in the family $\Gamma(|C|, |I|)$ if it has the same properties as described in (i), (ii), and (iii) above. The coefficient κ is then defined as

$$\kappa = \begin{cases} \dfrac{|C|}{|I|q^*} & \text{if } G^* \text{ exists,} \\ 1 & \text{otherwise,} \end{cases}$$

where q^* is the value of the coefficient $K = q$ for the grammar G^*. If G^* exists, then the efficiency of any grammar $G \in \Gamma(|C|, |I|)$ is given by

$$\text{Eff}(G) = \frac{q}{q^*}. \tag{18}$$

Based on (14), (16), and (17), it follows that

$$q = \frac{3}{11} \cdot \frac{7}{23} = \frac{21}{253} \tag{19}$$

for the Turkish PoS system and it remains to find q^* in the family $\Gamma(3, 4)$. This is much simpler than finding Q^*, which is needed in the previous formula (12). First, $|\Phi| = |I| = 4$ and $|B| = 2$ hold true for any grammar in $\Gamma(3, 4)$ satisfying the property (i). This implies that $\mu(\Phi) = 2$. Second, the greatest value of $\bar{Q}$, denoted by $\bar{Q}^*$, is $\bar{Q}^* = 1$ for any family of grammars $\Gamma(|C|, |I|)$ in which G^* exists. This is because the set R can be enlarged so that $\rho = \bar{\rho}$. Since there are several maximally efficient grammars in $\Gamma(3, 4)$, one of them being the grammar with $C = \{\text{H, a, m}\}$, it follows from (17) that

$$q^* = \frac{1}{2}, \tag{20}$$

which by (18) and (19) implies

$$\text{Eff} = \frac{42}{253} = 0.166$$

for the Turkish PoS system. This value is not that far from the previously calculated 0.112, but is found much more easily.

It should be pointed out that (13) is only a simplified version of the formula introduced in Vulanović and Ruff (2018), which contains two generalizations: a parameter that can be chosen by the user and a weighted version of the set-cardinality $|\cdot|$. The generalizations are not considered here for the sake of simplicity. The weights that can be used in $|\cdot|$ may somewhat complicate the search for q^*, but the procedure still remains simpler than finding Q^*, primarily because the value $\bar{Q}^* = 1$ is preserved. It is the search for the smallest value of $\mu(\Phi)$ that may get more involved, as the property that $\mu(\Phi)$ has a fixed value for all grammars in $\Gamma(3, 4)$ which satisfy (i), is not preserved when the weights are used. In fact, this is not a general property of all families $\Gamma(|C|, |I|)$ even when the weights are not used. For instance, in $\Gamma(2, 4)$, $\mu(\Phi)$, as defined in (13), has different values for the grammar with $C = \{V, \Lambda\}$ and the one with $C = \{H, M\}$ because the values of $|B|$ are not the same.

5 Conclusion

Natural languages are specific complex systems and, accordingly, their efficiency is measured in a special way. It is in fact the efficiency of their grammars that can be measured more easily. Although grammars are artificial descriptions of natural languages, they are easier to formalize and then the definition of grammar efficiency can be based on this formalization. In the preceding sections, it is presented how this can be done. Regardless of the specificities that natural languages have, it is shown that the initial approach to grammar efficiency can be fairly general and of the same kind as in other complex systems with identifiable and measurable inputs and outputs.

The formal grammar used here is similar to Dik's Functional Grammar, Dik (1997) (see Vulanović 2003 for more details), but it also has a regulated-rewriting component, Dassow and Păun (1989). Even though the approach taken in this paper is formal, the formalization is not carried out completely. Rather, some concepts are explained by examples, the Turkish part-of-speech system (PoS) being the main one. The model considered is relatively simple, which agrees with Miestamo's notion of local language complexity, Miestamo (2008). However, the same approach can be applied, in principle, to more complex models. On the other hand, even the simple model of the Turkish PoS system shows that it may be quite complicated to measure grammar efficiency using an earlier method from Vulanović (2003, 2007). This motivates the search for a new, simpler, method. The search is concluded here with the inclusion of a new result for measuring the degree of violation of the one-meaning–one-form principle, Vulanović and Ruff (2018). Both the old approach and the newly proposed one are presented and compared. The Turkish model shows that the two approaches give similar grammar-efficiency measures, but that the new one is indeed simpler. I should like to mention at this point that I have tested the new method on the grammars of simple transitive sentences, considered in Vulanović (2007). The results are similar, and in many cases even identical, to those of the old method. This justifies further investigations of the new grammar-efficiency formula, particularly since only the simplest version of the result from Vulanović and Ruff (2018) is considered here.

References

Anttila R (1972) An introduction to historical and comparative linguistics. Macmillan, New York

Dassow J, Păun G (1989) Regulated rewriting in formal language theory. Springer, New York

Dik SC (1997) The theory of functional grammar, part 1: the structure of the main clause. Mouton de Gruyter, Berlin

Hengeveld K (1992) Parts of speech. In Fortescue M, Harder P, Kristoffersen L (eds) Layered structure and reference in functional perspective. John Benjamins, Amsterdam/Philadelphia, pp 29–55

Hengeveld K, van Lier E (2010) Parts of speech and dependent clauses in functional discourse grammar. In Ansaldo U, Don J, Pfau R (eds) Parts of speech: empirical and theoretical advances. John Benjamins, Amsterdam/Philadelphia, pp 253–285

Hengeveld K, Rijkhoff J, Siewierska A (2004) Parts-of-speech systems and word order. J Linguistics 40:527–570

Miestamo M (2008) Grammatical complexity in a cross-linguistic perspective. In Miestamo M, Sinnemki K, Karlsson F (eds) Language complexity: typology, contact, change. Studies in language companion series, vol 94. John Benjamins, Amsterdam, pp 23–41

Vulanović R (1991) On measuring grammar efficiency and redundancy. Linguistic Analy 21(3–4):201–211

Vulanović R (1993) Word order and grammar efficiency. Theoret Linguistics 19(2–3):201–222

Vulanović R (2003) Grammar efficiency and complexity. Grammars 6:127–144

Vulanović R (2007) On measuring language complexity as relative to the conveyed linguistic information. SKY J Linguistics 20:399–427

Vulanović R (2008) A mathematical analysis of parts-of-speech systems. Glottometrics 17:51–65

Vulanović R, Ruff O (2018) Measuring the degree of violation of the one-meaning–one-form principle. In Wang L, Köhler R, Tuzzi A (eds) Structure, function and process in texts. RAM-Verlag, Lüdenscheid, pp 67–77

Self-Organizing Computational Efficiency in Quranic Grammar

Mahmoud Shokrollahi-Far

Abstract The existing knowledge-based and data-driven systems for Arabic morphological analysis are all suffering three main computational drawbacks, viz. efficiency, domain, and abstraction. Although the knowledge-based systems employ heavy lexical databases, they generate highly ambiguous tags. And to cover a new domain their lexicon should be costly modified. They also do not provide the linguistic abstraction preferred especially in computational linguistics. Similarly, the systems developed following a data-driven approach ignore the linguistic tractability for Arabic morphology and are highly dependent on big sizes of domain-specific training data. The source of these drawbacks may be traced in the morphological approach they employ in their knowledge base or in their training data. This chapter introduces regex morpho-syntax for Arabic, a highly efficient formalism originating from the basic grammatical rules developed for diacritizing the Quran fourteen centuries ago. The developed formalism is implemented in the knowledge base of Mobin morpho-syntactic parser and tagger. The achieved F-score of 0.967 for the computational effectiveness of the system as well as its significant comparative efficiency measured in terms of Kolmogorov complexity highlights the inherent computational efficiency in Quranic grammar.

Keywords Arabic language · Grammatical complexity · Computational natural language · Processing systems · Artificial Intelligence · Knowledge-based systems · Quranic grammar · Mobin parser and tagger

---

M. Shokrollahi-Far (✉)
University College of Nabi-Akram (UCNA), Tabriz, Iran

Tilburg University, Tilburg, Noord-Brabant, The Netherlands
e-mail: msf@ucna.ac.ir; m.shokrollahifar@uvt.nl

G. Y. Georgiev, M. Shokrollahi-Far (eds.), *Efficiency in Complex Systems*,
Springer Proceedings in Complexity, https://doi.org/10.1007/978-3-030-69288-9_8

1 Introduction

NLP systems are developed following three AI approaches, viz. (a) **knowledge based**, (b) **data driven/empirical**, and (c) **hybrid**. In the knowledge-based systems the required knowledge is formalized and embedded in the rules of the knowledge base of the systems. In contrast, in the data-driven approach the required knowledge for the system is explicitly or implicitly rooted in the data by which the machine learning (ML) system is trained. And the hybrid systems are a combination of the knowledge-based and the data-driven ones. Following any of the AI approaches, almost all NLP systems are more or less suffering three common drawbacks, viz. (1) **efficiency**, (2) **domain**, and (3) **abstraction**. Existing Arabic NLP systems are not an exception to these drawbacks.

1.1 Efficiency

At the side of the knowledge-based approach, the analysis in the systems relies on the linguistic rules represented in their "heavy lexicon" (Al-Othman and Al-Salman 2020) which is the main source of complexity and leads to the computational inefficiency in the system. Moreover, in their rules all the possible structures of the language are formalized to return all the possible analyses for a given input for further processing by humans or another NLP component (Soudi et al. 2007). Unfortunately, most systems do not dispose any decision criteria on how to rank or select between the alternative readings in a given domain or application. This results in high **ambiguity** in the systems and hence its inefficiency. And if the system employs some extra decision criteria in their lexicon to resolve the ambiguity, this would increase its complexity in its lexicon and hence its inefficiency.

Similar inefficiencies are witnessed in the data-driven approach. First, the performance of the ML schemes is directly affected by the size of the training data and the richness of the annotations in it. So, the bigger the size of the training data and the richer its annotations, the better the effectiveness in the system. This great dependence of the performance of the ML systems on a big size of annotated data is the main basis of their inefficiency. Second, the same ambiguity in the output of knowledge-based systems is usually seen in the output of some data-driven systems.

1.2 Domain

In the data-driven approach the ML schemes usually need their training data to be annotated in a way relevant to the application or domain for which the NLP system is developed (Soudi et al. 2007). The availability of such relevant annotated corpora is usually a serious challenge. In contrast, the knowledge-based systems are often assumed to be domain independent in a way that the linguistic knowledge could cover all possible structures and only these. However, this requires not only that all possible allowable structures are known by the linguist, but also that they can be formalized and implemented consistently preferable as nonredundant as possible (Wintner 2007). And this is a main reason why the existing knowledge-based systems are highly dependent on heavy lexical databases specified to certain domains.

1.3 Abstraction

The major benefit of the knowledge-based methods is that their rules and constraints are defined on a precise linguistic basis for recognizing and classifying the internal structure of words. However, many Arabic NLP systems are suffering the lack of abstraction and difficulties in incremental processing (Soudi et al. 2007). Similarly, most ML schemes do not let us follow their computing procedures heuristically, despite the fact that the precise linguistic research requires a sound formal basis.

Among the efficiency, domain, and abstraction drawbacks of the Arabic NLP systems it is their efficiency that is more considerable from computational point of view, especially in the knowledge-based systems. This is more considerable when there are few systems developed following a data-driven approach and most of the Arabic NLP systems are developed on a knowledge-based approach (Al-Othman and Al-Salman 2020). Reviewing such systems is the main concern of the next section.

2 Arabic Knowledge-Based NLP Systems

The Arabic NLP systems are mostly knowledge-based systems, relying on heavy lexical lists or databases and formal rules developed on a linguistic basis. For an elaborate survey of the reported systems developed till 2004 the reader may refer to al-Sughaiyer and al-Kharashi (2004) and for the systems since then till 2020 to al-Othman and al-Salman (2020). Even though the lexical databases in these systems are very similar, they have employed different underlying linguistic frameworks.

2.1 Linguistic Frameworks in Arabic NLP Systems

As common for the knowledge-based NLP systems, the basic aim when developing the similar systems for Arabic is providing a linguistically tractable account of Arabic morphology. The major linguistic frameworks employed in these systems are classified into four approaches in morphology, viz. (a) **root and pattern**, (b) **lexeme**,

(c) **stem**, and (d) **syllable** (Soudi et al. 2007) and (Al-Othman and Al-Salman 2020). Among these frameworks computational linguists have mostly accepted "root-and-pattern morphology" as their linguistic framework to model Arabic morphology. A good example is McCarthy's **auto-segmental prosodic** morphology (McCarthy 1981) that is reflected in most of the computational attempts, especially in the systems written within **finite-state morphology**. McCarthy's theory is discussed in Sect. 5.

2.2 *Some Arabic NLP Systems and Their Drawbacks*

Dichy and Farghaly (2007) developed the first famous system on stem-based morphology in which the focus of the representation was the stem, the operations with grammar, and the lexis specifications on the stem. And in separate lexicons there were the root, the pattern (or the template), and the vocalism of a word. Their work relied on two resources in Arabic NLP development, viz. (a) DIINAR.1 *Dictionnaire Informatisé de l'Arabe*, a comprehensive Arabic lexical resource of around 121,000 lemma entries, and (b) the lexical database and analyzers embedded in the SYSTRAN Arabic-English translator, a fully automatic transfer system. These heavy lexical databases are the main source of computational inefficiency in their system. Al-Othman and al-Salman (2020) have reviewed similar stem-based Arabic NLP systems, namely MORPHE tool reported in 2000 and 2001, MAGEAD and CALIMA tools in 2005 and 2006, ElixirFM and MORGEANA tools in 2007, MADA tool in 2009, AraComLex tool in 2011, and MADAMIRA tool in 2014.

The next famous Arabic stem-based system is the Buckwalter Arabic Morphological Analyzer (BAMA version 2.0) (Buckwalter 2004) employed at Linguistic Data Consortium, University of Pennsylvania to distribute its Arabic corpora in 2009. The system uses three Arabic-English lists for prefixes, suffixes, and stems that are supplemented by three morphological compatibility tables. Implemented in Perl script, the total size of the lists and the tables is nearly 4MB, the heaviest lexical database reported for Arabic NLP so far, which makes the system computationally inefficient. As another source of inefficiency, BAMA generates highly **ambiguous** outputs for the single input word. In the system, employing some dictionary lookup function, the input words are checked for the existence of the prefix, stem, and suffix into which the input word is first segmented. When all three components are found in their respective tables, it is then determined whether their respective morphological categories are compatible. The analysis is reported only if all three pairs are found in their respective tables, the three components are compatible, and the word is valid. And if a word returns no analysis, the orthography of the input string will be checked and a list of alternative spellings will be created.

BAMA was substituted in 2010 by Standard Arabic Morphological Analyzer (SAMA Version 3.1) which contains a lot of implementational differences, but the overall input/output compatibility is maintained. Simultaneously, to disambiguate the output of BAMA or SAMA automatically MADA+TOKAN was developed empirically on the Arabic corpora of LDC. Its accuracy on the basic morphological choice (including tokenization but excluding case, mood, and nunation) and on lemmatization is over 96% (Habash et al. 2009). MADA+TOKAN is a hybrid system developed by Habash et al. (2009), in which both knowledge-based and data-driven approaches are integrated. First the knowledge-based system MADA is employed to generate the tagged corpora, returning all possible analyses for a given input. Then the resulted solutions in the corpora are disambiguated employing TOKAN data-driven system. This is the dominant approach applied for the Arabic corpora distributed at LDC. The main drawback here is the fact that the disambiguation procedure is usually very costly, making it inefficient.

As the most famous system employing root-and-pattern morphology Beesley (1996a) developed ALPNET, one of the largest systems ever built for Arabic morphology and implemented using Xerox Finite-State Technology (XFST). The system employs three main components, viz. (a) a set of roots, (b) a set of templates, and (c) a lexicon of Arabic lexemes, which are the first source of inefficiency in the system. Besides, Beesley (1996b) had to restrain the set of vowels, due to the fact that most Arabic newswires are not vocalically **diacritized**. Moreover, since there was no other constraining condition for the readings in the output of ALPNET, its output was highly ambiguous, making the system highly more inefficient. Furthermore, the system generated Arabic verbs and nouns on the intersection of two sets, a set comprising the roots and another set comprising the patterns, which were retrieved from an existing big lexicon including Arabic lexemes. This would require the modification of the lexicon to extend the coverage of the system and to generalize it to new domains. Hence, the dependence of the system on a big lexicon not covering all domains encountered it with the second drawback of such systems. Finally, the abstraction problem of XFST propagated in ALPNET, echoing the third drawback of the existing knowledge-based systems for Arabic morphological analysis.

Employing Beesley's (Beesley 1996a) XFST, Dror et al. (2004) implemented a morphological tagger for the Quran, which uses three lexicons for the closed-class words, the noun base words, and the verb base ones containing roots and patterns of the words. The closed-class and the noun-based entries were constructed manually, but the verbal based ones were generated by concatenating all possible combinations of the roots and the patterns. Besides, a set of finite state-based rules check for the phonological transformations of the nouns and the verbs in the Quran. Then a morphological analyzer was implemented by compiling the existing finite-state toolbox into finite-state networks. The lexicons contain 2.5K entries for nouns, 100K entries for the verbs, about 50 rules for the nouns, and 300 rules for the verbs. The output is stored in a database for efficient searches based on a variety of keys. However, there might be many analyses per word, and the system is unable to resolve the resulted ambiguities, which make the system highly inefficient. To evaluate the performance of the system Dror et al. (2004) compared the results with a manually tagged fragment of the Quran, and reported 93% precision and 80% recall, but an average ambiguity of 1.15.

Better apprehension of the knowledge-based systems in Arabic NLP requires a deeper linguistic insight into Arabic morpho-syntax which is described as follows in Sects. 3, 4, and 5.

3 Arabic Orthography

Arabic alphabet consists of 28 **consonantal** letters, without any letters for its vowels, written in a right-to-left cursive script. In its orthography many of the consonants look similar in their basic shape being distinguished by **dots**, as the integral parts of a letter, above or below their central part. For example, in the consonants ح *h*, خ *kh*, and ج *j* in Arabic together with چ *ch* in Persian it is the dots that are distinguishing the consonants from one another. Considering the dots as **diacritics** for the consonantal letters, there are a few other diacritics which represent Arabic **vowels** and may appear "optionally" in the script.

So, in the Arabic script the word comprises mainly the consonantal letters and the vocalic diacritics may be missing from the word. In these cases there would be different realizations for the word, because of the morphological fact that in an Arabic word "the same string of the consonantal letters may accept different vocalic diacritics." Example [1] is an Arabic lexeme consisting of merely consonantal letters from which the diacritics are missing and may be realized as one of the five different words in the examples [1a,b,c,e,f]. A special vocalic case in Arabic orthography is سكون *suku:n* "vowel stop" that indicates the potential position for a vowel and is represented by the optional diacritics ◌ْ λ, as in Example [1g].

Example [1] كتب KTB 'write' without any vocalic diacritics
[1a] كِتَاب *kita:b* 'book' with two vocalic diacritics ◌ِ *i* and ا *a:*
[1b] كُتُب *kutub* 'books' with two vocalic diacritics that are both ◌ُ *u*
[1c] كَاتِب *ka:tib* 'writer' with two vocalic diacritics ا *a:* and ◌ِ *i*
[1d] كُتَّاب *kutta:b* 'writers' with two vocalic diacritics ◌ُ *u* and ا *a:*
[1e] كَتَبَ *kataba* 'he wrote' with three vocalic diacritics that are all ◌َ *a*
[1f] كُتِبَ *kutiba* 'it was written' with three vocalic diacritics ◌ُ *u*, ◌ِ *i* and ◌َ *a*
[1g] مَكْتَب *makλtab* 'class' with one auxiliary consonant *m* and two vocalic diacritics that are both ◌َ *a*

Example [1] is the lexeme, Example [1a] its singular noun, Example [1b] its plural noun, Example [1c] its singular agent noun, Example [1d] its plural agent noun, Example [1e] its active simple past singular masculine verb, Example [1f] its passive simple past singular masculine verb, and Example [1g] its singular locative noun. The different realizations in examples [1a,b,c,d,e,f,g] of the lexeme in Example [1] clearly illustrate "the morpho-syntactic significance of the vocalic diacritics in the Arabic words." So, in Arabic, as in all Semitic languages, the consonantal lexeme carries the main semantic load, whereas the vowels and the auxiliary consonants provide the rest of the morphological information (Versteegh 1997).

When these vowels are missing from the written word, it may cause so many "ambiguities" for the less experienced readers of the Arabic texts, especially when even the dots are missing from the consonants. This was the real problem for the Quran in its beginning time with the newcomers to Islam who spoke foreign languages different from Arabic. Actually, the Holy Quran was orally revealed to Prophet Muhammad more than fourteen centuries ago, and His followers first transcribed the book in the Arabic primitive script which lacked any diacritics and dots (Khorramshahi 1994; Khorramshahi 1999). To handle the problem, after a short while the required dots and vocalic diacritics, collectively called اعجام *I'ja:m*, were added to the text of the Quran in a very systematic morpho-syntactic annotation procedure.

This was a sophisticated formal task which required developing a complete "grammar" for the language of the Quran at that time. This was accomplished shortly after the holy book was transcribed when Ali Ibn Abi Talib (13/9/601–29/1/661), the fourth Caliph of Islam (656 to 661) and the cousin and the son-in-law of Prophet Mohammad, founded the first two basic grammatical rules for Arabic that are presented in Definition 1. He asked his poet companion Abu al-Aswad al-Du'ali (603–689) to develop the first two basic grammatical rules into a complete grammar for the language (Versteegh 1997).

4 Typology of Arabic Morphology

Katamba (1993) categorized the morphological typology of languages into the following five types:

(a) **Analytic** (also **isolating**), such as English and Chinese, in which each free morpheme normally occurs as a word in isolation.
(b) **Agglutinating** (also **agglutinative**), such as Turkish, in which most morphemes are bound and there is normally a more or less one-to-one matching of morphemes with morphs.
(c) **Inflecting** (also **synthetic** or **fusional**), such as Latin and Arabic, in which words normally consist of several morphemes, but a single morph may represent several morphemes simultaneously.
(d) **Incorporating** (also **polysynthetic**), such as Arabic, in which one word may include a verb and its object, equivalent to a whole sentence in English containing several words.
(e) **Infixing**, that is typical of Semitic languages like Arabic and Hebrew, in which much inflection involves infixing vowels in a root of consonants.

So, Arabic is not only an inflecting and incorporating but more differently an infixing language. More formally speaking, in categories (a)–(d) words are generated through a concatenation process in which some morphemes are concatenated both to derive a new word and to inflect it. In contrast, in Semitic languages, as defined in (e), the stem of base words is derived in a non-concatenative process. Modeling this derivation process is the main challenging concern of the different linguistic frameworks listed in Sect. 2.1, and is going to be elaborated on in the subsequent sections.

After the stem of the Arabic base word is derived in a non-concatenative infixing process, some prefix particles and/or suffix pronouns may be concatenated to the base word as proclitic and enclitic, respectively. For example, let us consider the Quranic polysynthetic word in Example [2]. The base word of this complex word is in Example [2a], the stem of the base word is in Example [2b], the inflecting prefix for the stem is in Example [2c], the two proclitic particles to the base word are in Example [2d], and its two enclitic pronouns are in Example [2e].

Example [2] فَسَيَكْفِيكَهُمْ *fasayakfi:kahum' So he will suffice you against them'*
[2a] يَكْفِى *yakfi:*'he suffices' in which the stem is
[2b] كْفِى *kfi:*'to suffice'
[2c] ى *ya*
[2d] فَ *fa* 'so' and سَ *sa* 'will'
[2e] كَ *ka* 'you' and هُمْ *hum* 'them'

5 Prosodic Morphology and Its Drawback

As explained in Sect. 2.1, the most accepted linguistic framework reflected in most of the computational attempts to model Arabic morphology is **root-and-pattern** morphology. And a good example of it is McCarthy's theory of **prosodic** morphology (also **template** morphology) that is heavily based on **auto-segmental phonology**, the theoretical formalism of the generative phonology model initially used to describe tone (Katamba 1993). In this theory, which incorporates the **morpheme tier hypothesis**, phonological representations consist of several independent, parallel **tiers** (i.e., levels of representation) that are not isolated from each other. In his prosodic approach to Arabic morphology McCarthy (1981) noted the similarity between how the vowels are morphologically spread over the consonantal roots and how the phonological prosodies like tones are spread (Sproat 1992).

He hypothesized that, at the underlying level of the representation in the lexicon, the Arabic verb, e.g., كُتِبَ *kutiba* "it was written" in Example [3] (the same word in Example [1f]), has elements arranged on three independent tiers, viz. (a) the **vocalic melody tier** *uia*, (b) the **skeletal tier** (or **prosodic template tier** or the **CV tier**) CVCVCV, and (c) the **root tier** (also called **consonantal** tier) ك ت ب *ktb* "write" (cf. Katamba 1993).

In the **morpheme tier hypothesis** a morpheme tier is represented by μ. So, in the word مَكْتَبٌ *maktabun* "a class" (nominative) in Example [4] there are five tiers, viz. (a) the root tier ك ت ب *ktb* "write" (μ_1), (b) the skeletal tier CVCCVCVC, (c)

Example [3]

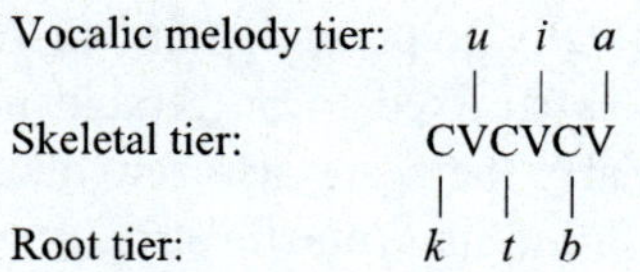

the vocalic melody tier *a* (μ_2), (d) the nominative morpheme tier ◌ٌ *un* "indicating nominative" (μ_4), and (e) the locative morpheme tier مَ *ma-* "indicating location" (μ_3) (Katamba 1993):

Example [4]

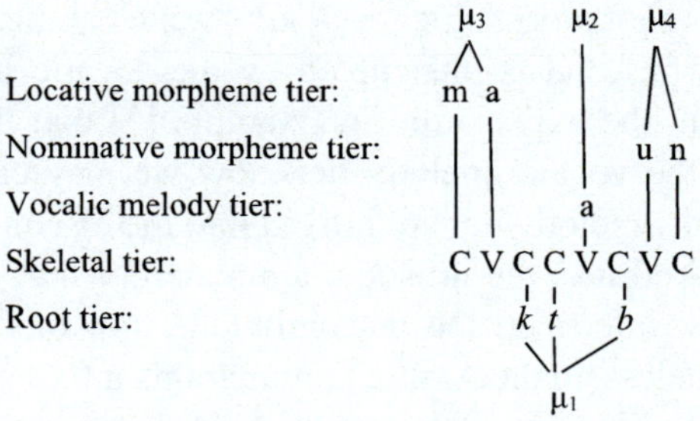

However, McCarthy (1981) overlooks the fact that the prefix morpheme مَ *ma-* "indicating location" (μ_3) is valid only if it precedes a word in which the discontinuous vocalic *a* is infixed into a root morpheme like كتب *ktb* "write" (μ_1). So, the locative morpheme tier should be merged into the vocalic melody tier, setting up a more comprehensive tier in μ_2. Consequently, the five-tier structure in Example [4] should be reduced into a four-tier one, as in Example [5]:

Example [5]

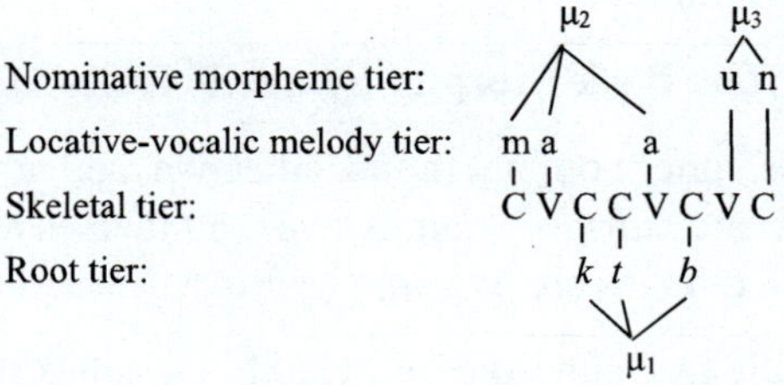

One morphologically important point in Example [5] is that the nominative morpheme tier μ_3 is suffixed to an Arabic noun only after it is employed in a sentence, and similarly there are also few other prefixes and suffixes that may be affixed to such a word in its syntactic structure in the sentence. Moreover, prosodic morphology expresses that in the non-concatenative morphological systems, in which words are not necessarily made up of successive sequences of morphemes, all the sounds that represent the same morpheme are not necessarily adjacent to each other and it is common to find root morphemes as **discontinuous morphemes** interrupted by infixes (McCarthy 1981).

In this sense, in the noun مَكْتَبٌ *maktabun* "a class" (nominative) there is the discontinuous root morpheme كتب *ktb* "write" at the root tier carrying the sematic load of the word that is interrupted by the vocalic infix *a* at the vocalic melody tier. Following the explanation in Example [5] that the locative morpheme tier is merged into the vocalic melody tier, now we may conclude that *ma* and *a* at the locative-vocalic melody tier are infixed into the discontinuous root morpheme كتب *ktb* "write." And it is the prosodic template tier that arranges the infixing process of the vocalic infixes in the discontinuous root morpheme at the morphological level. Nevertheless, in the Arabic grammar this infixing process is viewed differently making it more efficient, elaborated on as follows in Sect. 6. This discussion lets us to move from McCarthy's prosodic morphology to **regex morpho-syntax**.

6 Regex Morpho-Syntax in Arabic

In the concluding part of Sect. 3 it was explained that Ali Ibn Abi Talib fourteen centuries ago founded the first two most fundamental linguistic basis in Arabic as follows in Definition 1.

Definition 1: The Basic Morpho-Syntactic Rules in Arabic Grammar

(a) Words are either nominal (noun, adjective, and adverb), verb, or particle.
(b) All agents are *marfu:ʕ* "nominative"; all themes are *mansu:b* "accusative"; and all modified nouns are *majru:r* "genitive."

The first rule in Definition 1 efficiently classifies the parts of speech for Arabic words, and the second rule classifies Arabic nominal cases together with the governing operators for the cases. A significant grammatical point is that the three Arabic nominal cases are not named for their "case" functions, but based on the Arabic إِعْرَاب *iʕra:b* "declensional system." This Arabic morpho-syntactic system was the most essential innovative systematization in Arabic grammar founded by Sibawayhi (760–796), an Iranian intimate grammarian fellow of Abu al-Aswad al-Du'ali (Versteegh 1997), and is described as follows in Definition 2.

Definition 2: Arabic Declensional System To indicate the nominal case or the verbal mood forms in Arabic the declension of the noun or verb takes place by the inflection of the vowel of its exponent, which for the nominative, the accusative, and the genitive cases includes (a) *-u*, *-un*, or *-u:na*; (b) *-a*, *-an*, or *-i:n*; and (c) *-i*, *-in*, or *-i:n* nominal suffixes, respectively, and for the indicative, the subjunctive and the jussive moods include (a) *-u*, *-u:na*, *-i:na*, *-a:ni*, or *-na*; (b) *-i*; and (c) zero-vowel verbal suffixes, respectively.

In Definition 2 the vowels differentiating the declension marks in the nominative, the accusative, and the genitive cases are /u:/, /a/, and /i/, respectively. This is illustrated as follows in Example [6]:

Example [6] نَصَرَ حَسَنٌ زَيدَ
nasara Hasanu Zayda.
nasara ^'helped' VERB^ *Hasan*^SUBJECT^ _*u*^NOMINATIVE MARK^ *Zayd*^OBJECT^ _*a*^ACCUSATIVE MARK^

So, the Arabic "canonical theory of grammar" is divided into two parts, namely syntax and morpho-phonology (Bohas et al. 1990). The core of Arabic syntax is the use of case and mood declension marks together with their distribution in a phrase, comprising a "theory of government" being performed always in a noun phrase (NP) or a verb phrase (VP) as follows in Definition 3.

Definition 3: Theory of Government in Arabic Syntax In an NP or a VP a declension mark is determined on the exponent of the head word of the phrase as the operand by an "operating element/governing operator" that normally occurs before its operand.

For instance, in Example [6] the verb نَصَرَ *nasara* "helped" is the governing operator that determines the nominative mark *-u* in the subject noun *Hasan* and the accusative mark *-a* in the object noun *Zayd*. This leads us to the **governing context** for the Arabic nouns and verbs as follows in Definition 4.

Definition 4: Governing Context of Arabic Nouns and Verbs The finite set *G*= {(*g*, *-t*)| *g* is a "governing operator" and *-t* is an exponent in the operand} comprises the governing context of Arabic nouns and verbs.

At this point we can present the very noticeable formal definition of Arabic verbs and nouns. Foremost it should be noted that since thirteen centuries ago Arabic grammar has represented the vocalic melody and skeletal tiers of prosodic morphology employing especial **regular expressions** (**regex**) as follows in Definition 5.

Definition 5: Arabic Derivational Regular Expressions In the pattern فِعْل *fiςl* "verb" the letters ف *f*, ع *ς*, and ل *l* represent, respectively, the first, the second, and the third root radical consonants infixed by some permutations of the three vocalic diacritics ◌ُ *u*, ◌ِ *i*, and ◌َ *a* and possibly some auxiliary consonants, building up the valid regular expressions (regex) for Arabic derivational words.

Example [7] presents the regexes matching the words derived on the root morpheme كتب *ktb* "write."

Moreover, at the morpho-syntactic level, it is the nominal regex such as مَفْعُول *mafςu:l* "maCCu:C" that takes the case marker suffixes and yields the nominative

Example [7] Verbal regex فُعِل *fuʕila* 'CuCiCa' matches كُتِبَ *kutiba* 'it was written'
Verbal regex فَعَل *faʕala* 'CaCaCa' matches كَتَبَ *kataba* 'he wrote'
Nominal regex فُعُل *fuʕul* 'CuCuC' matches كُتُب *kutub* 'books'
Nominal regex مَفعُول *mafʕu:l* 'maCCu:C' matches مَكتُوب *maktu:b* 'written text'

nominal regex مَفْعُولٌ *mafʕu:lun* "maCCu:Cun" matching the nouns like مَكْتُوبٌ *maktu:bun* "a written text" (nominative). Following Definition 5, among the Arabic parts of speech it should significantly be noted that all Arabic verbs are recognized in a "regex-matching" process as follows in Definition 6.

Definition 6: Formal Definition of Arabic Verbs An Arabic verb is any word that matches one of the valid perfective or imperfective, trilateral or quadrilateral, مُجَرَّد *mujarrad* "basic" or مَزيد *mazi:d* "augmented" verbal regexes listed in Table 1.

Note 1: In Table 1, the regexes in the rows I-1 to I-3 and those in the rows II to X are Arabic trilateral مُجَرَّد *mujarrad* "basic" and مَزيد *mazi:d* "augmented" verbal regexes, respectively, that consist of three radical consonantal slots and match any word with a trilateral root morpheme like كـتـب *ktb* "write." Likewise, the regexes in the rows XI and XII are Arabic quadrilateral مُجَرَّد *mujarrad* "basic" and مَزيد *mazi:d* "augmented" verbal regexes, respectively, that consist of four radical consonantal slots and match any word with quadrilateral root morpheme like دح رج *dhrj* "roll." As an example, the trilateral مُجَرَّد *mujarrad* "basic" verbal regex فعل *faʕala* "CaCaCa" matches the verb كَتَبَ *kataba* "he wrote," and the trilateral مَزيد *mazi:d* "augmented" verbal regex فَعَّل *faʕʕala* "CaCCaCa" matches the verb كَتَّبَ *kattaba* "he caused to write," both having the trilateral root morpheme كـتـب *ktb* "write."

Definition 7:Arabic Derivational Nominal Regexes Arabic derivational nouns are any word that matches one of the valid trilateral or quadrilateral, مُجَرَّد *mujarrad* "basic" or مَزيد *mazi:d* "augmented" verbal based nominal regexes listed in Table 1, or one of the nominal regexes فاعِل *fa:ʕil* "Ca:CiC" for nominative nouns, مَفعول *mafʕu:l* "maCCu:C" for accusative nouns, مَفعَل *mafʕal* "maCCaC" for locative nouns, and مَفعِل *mafʕil* "maCCiC" for temporal nouns.

For example, the trilateral verbal based مَزيد *mazi:d* "augmented" nominal regex مُفاعَلة *mufa:ʕalah* "muCa:CaCah" matches the noun مُكاتَبة *muka:tabah* "writing to each other," with the trilateral root morpheme كـتـب *ktb* "write."

The regexes for the verbal bases in Table 1 comprise two strings, viz. one **derivational stem** and one **inflectional affix**. In general, there are 14 verbal inflectional suffixes for the perfective stem and 14 verbal circumfixes for the imperfective stem, presented in bold face in Table 2. For example, in the perfective verbal regex فَعَلَ *faʕala* "CaCaCa" the inflectional diacritic suffix for person is ـَ *-a*, and in the imperfective verbal regex يَفْعَلُ *yafʕulu* "yaCCuCu" the inflectional circumfix for person is ي *ya* + ـُ *–u*. These affixes indicate perfective and imperfective aspects; sing (singular), dual, or pl (plural) number; mas (masculine) or fem (feminine) gender; and 1st, 2nd, or 3rd person.

Table 1 Arabic verbal and verbal based nominal regexes

	Perfective verbal	Imperfective verbal	Verbal based nominal
I-1	فَعَلَ *faʕala* "CaCaCa"	يَفْعَلُ *yafʕalu* "yaCCaCu"	–
		يَفْعِلُ *yafʕilu* "yaCCiCu"	–
		يَفْعُلُ *yafʕulu* "yaCCuCu"	–
I-2	فَعِلَ *faʕila* "CaCiCa"	يَفْعَلُ *yafʕalu* "yaCCaCu"	–
		يَفْعِلُ *yafʕilu* "yaCCiCu"	–
I-3	فَعُلَ *faʕula* "CaCuCa"	يَفْعُلُ *yafʕulu* "yaCCuCu"	–
II	فَعَّلَ *faʕʕala* "CaCCaCa"	يُفَعِّلُ *yufaʕʕilu* "yuCaCCiCu"	تَفْعِيل *tafʕi:l* "taCCi:C" تَفْعِلَه *tafʕilah* "taCCiCal"
III	فَاعَلَ *fa:ʕala* "Ca:CaCa"	يُفَاعِلُ *yufa:ʕilu* "yuCa:CiCu"	مُفَاعَلَه *mufa:ʕalah* "muCa:CaCah" فِعَال *fiʕal* "CiCaC"
IV	أَفْعَلَ *afʕala* "aCCaCa"	يُفْعِلُ *yufʕilu* "yuCCiCu"	إِفْعَال *ifʕa:l* "iCCa:C"
V	تَفَعَّلَ *tafaʕʕala* "taCaCCaCa"	يَتَفَعَّلُ *yatafaʕʕalu* "yataCaCCaCu"	تَفَعُّل *tafaʕʕul* "taCaCCuC"
VI	تَفَاعَلَ *tafa:ʕala* "taCa:CaCa"	يَتَفَاعَلُ *yatafa:ʕalu* "yataCa:CaCu"	تَفَاعُل *tafa:ʕul* "taCa:CuC"
VII	اِنْفَعَلَ *infaʕala* "inCaCaCa"	يَنْفَعِلُ *yanfaʕilu* "yanCaCiCu"	اِنْفِعَال *infiʕa:l* "inCiCa:C"
VIII	اِفْتَعَلَ *iftaʕala* "iCtaCaCa"	يَفْتَعِلُ *yaftaʕilu* "yaCtaCiCu"	اِفْتِعَال *iftiʕa:l* "iCtiCa:C"
IX	اِفْعَلَلَ *ifʕalala* "iCCaCaCa"	يَفْعَلِلُ *yafʕalilu* "yaCCaCiCu"	اِفْعِلَال *ifʕila:l* "iCCiCa:C"
X	اِسْتَفْعَلَ *istafʕala* "istaCCaCa"	يَسْتَفْعِلُ *yastafʕilu* "yastaCCiCu"	اِسْتِفْعَال *istifʕa:l* "istiCCa:C"
XI	فَعْلَلَ *faʕlala* "CaCCaCa"	يُفَعْلِلُ *yufaʕlilu* "yuCaCCiCu"	تَفَعْلُل *tafaʕli:l* "taCaCCi:C"
XII	تَفَعْلَلَ *tafaʕlala* "taCaCCaCa"	يَتَفَعْلَلُ *yatafaʕlalu* "yataCaCCaCu"	تَفَعْلُل *tafaʕli:l* "taCaCCi:C"

Table 2 Inflective affixes for Arabic verbs

	Number, gender, person	Perfective suffixes	Imperfective circumfixes
1	Sing, mas, 3rd	فَعَلَ ***faςala*** "CaCaCa"	يَفْعَلُ ***yafςalu*** "**ya**CCaC**u**"
2	Dual, mas, 3rd	فَعَلَا ***faςala:*** "CaCaC**a:**"	يَفْعَلَانِ ***yafςala:ni*** "**ya**CCaC**a:ni**"
3	Pl, mas, 3rd	فَعَلُوا ***faςalu:*** "CaCaC**u:**"	يَفْعَلُونَ ***yafςalu:na*** "**ya**CCaC**u:na**"
4	Sing, fem, 3rd	فَعَلَتْ ***faςalat*** "CaCaC**at**"	تَفْعَلُ ***tafςalu*** "**ta**CCaC**u**"
5	Dual, fem, 3rd	فَعَلَتَا ***faςalata:*** "CaCaC**ata:**"	تَفْعَلَانِ ***tafςala:ni*** "**ta**CCaC**a:ni**"
6	Pl, fem, 3rd	فَعَلْنَ ***faςalna*** "CaCaC**na**"	يَفْعَلْنَ ***yafςalna*** "**ya**CCaC**na**"
7	Sing, mas, 2nd	فَعَلْتَ ***faςalta*** "CaCaC**ta**"	تَفْعَلُ ***tafςalu*** "**ta**CCaC**u**"
8	Dual, mas, 2nd	فَعَلْتُمَا ***faςaltuma:*** "CaCaC**tuma:**"	تَفْعَلَانِ ***tafςala:ni*** "**ta**CCaC**a:ni**"
9	Pl, mas, 2nd	فَعَلْتُمْ ***faςaltum*** "CaCaC**tum**"	تَفْعَلُونَ ***tafςalu:na*** "**ta**CCaC**u:na**"
10	Sing, fem, 2nd	فَعَلْتِ ***faςalti*** "CaCaC**ti**"	تَفْعَلِينَ ***tafςali:na*** "**ta**CCaC**i:na**"
11	Dual, fem, 2nd	فَعَلْتُمَا ***faςaltuma:*** "CaCaC**tuma:**"	تَفْعَلَانِ ***tafςala:ni*** "**ta**CCaC**a:ni**"
12	Pl, fem, 2nd	فَعَلْتُنَّ ***faςaltunna*** "CaCaC**tunna**"	تَفْعَلْنَ ***tafςalna*** "**ta**CCaC**na**"
13	Sing, 1st	فَعَلْتُ ***faςaltu*** "CaCaC**tu**"	أَفْعَلُ ***afςalu*** "**a**CCaC**u**"
14	Pl, 1st	فَعَلْنَا ***faςalna:*** "CaCaC**na:**"	نَفْعَلُ ***nafςalu*** "**na**CCaC**u**"

So, similar to nominal governing contexts, another contextual grammatical set is defined for Arabic imperfective verbal regexes, as follows in Definition 8.

Definition 8: Circumfixes for Arabic Imperfective Verbal Regexes Let L_1, Σ_1, Σ_{P1}, and Σ_{S1} be disjoint finite sets of all imperfective verbal regexes, all imperfective verb stems, all imperfective verbal inflectional circumfixes, and all indicative verbal suffixes, respectively, in Arabic; let "+" mean "to the power zero or one"; then

$$L_1 = \left\{ r w_1 s^{+} t \middle| \, w_1 \in \Sigma_1, r \in \Sigma_{P1}, s \in \Sigma_{P1}, t \in \Sigma_{S1} \right\}$$

The verbal regexes are in **active** voice. Converting the voice from active into passive in the perfective and imperfective verbal regexes requires a vocalic transformation as follows in Definitions 9 and 10.

Definition 9: Transforming Perfective Verbal Regexes from Active into Passive To transform the voice of the perfective verbal regexes from active into passive the vowel following the second radical consonant is transformed into the vowel ِ *i*, and all the vowels preceding it are transformed into the vowel ُ *u*.

This voice transformation results in verbal regexes in passive voice. For example, the active perfective verbal regex إسْتَفْعَلَ *istafςala* "istaCCaCa" is transformed into passive أُسْتُفْعِلَ *ustufςila* "ustuCCiCa" regex.

Definition 10: Transforming Imperfective Verbal Regexes from Active into Passive To transform the voice of the imperfective verbal regexes from active into passive the vowel following the second radical consonant in the word is transformed into the vowel َ *a*, and the first vowel in the prefix is transformed into the vowel ُ *u*.

For example, the active imperfective regex يَسْتَفْعِلُ *yastafςilu* "yastaCCiCu" is transformed into passive يُسْتَفْعَلُ *yustafςalu* "yustaCCaCu" regex. There are similar transformations converting an imperfective verbal regex into imperative as follows in Definition 11.

Definition 11: Transforming Imperfective Verbs into Imperative To convert the imperfective verbal regexes into imperative the transformation is as follows.

(a) In the 1st and 3rd person, the proclitic particle لِ *li-* is prefixed to the verb.
(b) In the 2nd person, the inflective prefix of the verb is omitted; then if the first consonant is not followed by any vowel, the vowel ◌ِ *i* will precede the consonant, or the vowel ◌ُ *u* if the second radical consonant is followed by the vowel ◌ُ *u*.
(c) In the inflective suffix there is the omission of the final vowel ◌ُ *u* or the ending نَ *–na* except in the 2nd plural feminine.

For example, employing the root morpheme كتب *ktb* "write," for the imperfective 1st person singular verb أَكْتُبُ *aktubu* "I write" the imperative verb is لأَكْتُبْ *liaktub* "I must write," and for the imperfective 2nd person dual verb تَكْتُبانِ *taktuba:ni* "you two write" it is أُكْتُبا *uktuba:* "you two must write."

The next contextual grammatical set concerns how an Arabic base word may incorporate some **proclitics** and **enclitics**, e.g., examples [2d] and [2e], respectively. The general incorporating structure of an Arabic word is represented as follows in Definition 12.

Definition 12: Arabic Clitics Let L, Σ, Σ_{P1}, Σ_{P2}, Σ_{S1}, Σ_{S2}, and Σ_{S3} be disjoint finite sets of all Arabic words, all free morphemic words, all primary proclitic particles, all secondary proclitic particles, all primary enclitic pronouns, all secondary enclitic pronouns, and all enclitic particles, respectively, in Arabic; let "+" mean "to the power zero or one"; then

$$L=\{u^+v^+wx^+y^+z^+ \mid u \in \Sigma_{P1},\ v \in \Sigma_{P2},\ w \in \Sigma,\ x \in \Sigma_{S1},\ y \in \Sigma_{S2},\ z \in \Sigma_{S3}\}$$

Finally, employing the Arabic active trilateral imperfective verbal regexes and merging the sets already defined, Definition 13 introduces a regex for the first group of Arabic active trilateral imperfective verbal regexes.

Definition 13: A Group of Arabic Active Trilateral Imperfective Verbal Regex Let L', Σ_C, and Σ_V be disjoint finite sets for all active trilateral imperfective verbal regex يَفْعَلُ *yafςalu* "yaCCaCu," all the consonants, and all the vowels, respectively, in Arabic, and using the sets defined in Definitions 4, 8, and 12, then

$$L' = \left\{u^+v^+r\text{CCVC}s^+tx^+y^+z^+ \middle|\ \text{C} \in \Sigma_C, \text{V} \in \Sigma_V\right\}$$

Note 2: Based on all the verbal regexes in Table 1, a number of similar sets of regexes are developed for the verbs that are passive imperfective and imperative, as well as active and passive perfective. Similar sets of regexes are developed for the nominal regexes defined in Definition 7. All these regexes amount to nearly two thousands, and complementary to them there are the sets for Arabic non-governing particles, non-incorporating pronouns and demonstratives, and numerals. These regexes represent all the NPs and VPs in Arabic.

The nonformal and basic forms of the regexes developed here are brought about benefiting from el-Dahdah (1999), Muhammadi (2002), and al-Shartuti (1992). The collection of all these regexes comprises the main morpho-syntactic knowledge base of Mobin morpho-syntactic parser and tagger (Shokrollahi-Far 2012) to be explained in the next section.

7 Mobin Knowledge-Based Parser and Tagger

In order to overcome the defects of the Arabic knowledge-based systems discussed in Sect. 2.2, the Mobin knowledge-based parser and tagger is designed and implemented of which the first version was reported by Shokrollahi-Far et al. (2007b) as well as Shokrollahi-Far and Saraei (2007). Mobin bootstraps morpho-syntactic tags for every entry word in the input text in one-step parsing. All grammatically eligible regular expressions formalized in Definition 13 and noted in Note 2 are implemented as the knowledge base of Mobin using Perl scripting language. The total size of the source program of Mobin in a text file is less than 500K bytes.

On top of this knowledge base there are some pattern-matching functions that search through the vowelized input text, hence requiring the input text to be diacritized. In this search the functions first scan the input string of characters from space to space. Then calling one by one the regular expressions in the knowledge base, they search for the pattern which matches the input string. Based on the pattern matched, the morphemes in the input string are (a) mapped to the relevant attribute-value pairs in the morpho-syntactic tag set in Table 3 and (b) tagged using the appropriate tags in this tag set. The elaborate tag set is presented in Table 5 in the appendix of this chapter.

Table 3 Concise tag set of Mobin

Constraint	Attribute
Word segment	Affixation, part of speech, declination
Common for verb and noun	Alternation, generation, person, number, gender, case, root, structure, template, stem extra particle
Verb	Mood, voice, aspect, time
Noun	Derivation, derivation type
Particle	Specificity, operation, semantic

Table 4 Evaluation results of Mobin

Task	Average *F*-score
Tokenization	0.997
POS tagging	0.985
Morphological parsing	0.964
The whole system	0.967

Mobin outputs an xml file in which the words from space to space in the input text file are the main entries. The verb in Example [2] is parsed employing Mobin and tagged as in Example [8] as well as an Arabic noun in Example [9] that are presented in the appendix of this chapter. The abbreviations in the tags are also presented in Table 5. Mobin was automatically evaluated on the manually edited output of Mobin on the first chapter of the Holy Quran, occupying one-tenth of the book (Shokrollahi-Far et al. 2007a). The results are presented in Table 4 (cf. Shokrollahi-Far 2012).

8 Conclusions

The knowledge-based systems for Arabic morphological analysis are developed in order to provide a linguistically tractable account of Arabic morphology (Soudi et al. 2007). This is partially achieved in the existing systems for Arabic NLP compromising on efficiency, domain, and abstraction. First, although they employ heavy lexical databases, they generate highly ambiguous tags. Second, to cover a new domain their lexicon should be modified, which is a costly and exhausting task. And third, they have not provided the linguistic abstraction preferred especially in computational linguistics. The Arabic NLP systems developed following a data-driven approach to Arabic morphology are also suffering similar inefficiencies. Not only has this approach ignored the linguistic tractability for Arabic morphology, but also it is highly dependent on big sizes of domain-specific training data.

To resolve the drawbacks in the existing systems for Arabic NLP, **regex morpho-syntax** is developed and implemented in Mobin as a knowledge-based morpho-syntactic parser and tagger relying on the grammar of Arabic basically originated from systematically diacritizing the Quran fourteen centuries ago. Mobin effectively and deterministically (hence efficiently) bootstraps the grammatical tags in vocalically diacritized Arabic texts, without the need for any lexical databases of lexemes, roots, and stems. In terms of domain, the approach enables Mobin to be applied for tagging Arabic texts in any domain, say Islamic texts (see Shokrollahi-Far et al. 2009), without the need for any modification in the system. In terms of effectiveness, the evaluation results of the present version of Mobin (F-measure of 96%, on average) may be compared with those of the morphological tagger developed for the Quran at Haifa University (F-measure of 86%). Considering the evaluation of Mobin in POS tagging its F-measure is 98.5%, which is comparable to the reported recognition rate of 96% for the POS tagger in (Roth et al. 2008).

Nevertheless, for a more reliable evaluation on the domain coverage of Mobin, it should be evaluated on some other gold standards belonging to other domains and registers, say newswire, as well.

Moreover, where in the existing knowledge-based systems certain heavy lexicons are commonly used to let them generate all eligible solutions for any input string, Mobin acts independent of any such lexicons, and this does not hinder its ability to generate tags unambiguously, and hence very efficiently. This means that the regular expressions in its knowledge base are sufficient for modeling Arabic morpho-syntax, without requiring the kind of the rules employed in Beesley's (Beesley 1996a) system for two-level morphology developed for the analysis and generation of morphologically complex languages (cf. Karttunen and Beesley 2001). Moreover, there is no need in Mobin to compile its rules. So, the computational cost of Mobin is comparatively low that is another efficiency advantage for Mobin.

The most significant efficiency advantage of Mobin is the size of its source program that is highly competitive compared with the source size of similar systems, e.g., BAMA. The very low source size of the system makes it applicable for research on "the problem of optimal coding for a text" that is elaborated by Benedetto et al. (2002). As they explain, Shannon (1948) discovered a limit to the possibility of encoding a given sequence, which is "the entropy of the sequence" defined by Chaitin-Kolmogorov as "the length (in bits) of the smallest program which produces as output the string." As Benedetto et al. (2002) discussed, although it is impossible to find such a program, the file compressors or zippers are explicitly conceived to approach this theoretical limit. Although Mobin just parses the input string and does not generate a string yet (the way that ALPNET works), the very small size of its source program makes Mobin computationally highly efficient, hence the regex morpho-syntax in its knowledge base.

Furthermore, when parsing an Arabic text, the morpho-syntactic regexes are comprehensive and constraining enough to eliminate the need in Mobin for any supplementary lexicon. For instance, the roots in the stems of the bases of the words are extracted from the text itself without the need for any supplementary lexicon for Arabic roots. In this way, not only the abstraction drawback of ALPNET is overcome in Mobin, but also there is no need in Mobin to be extended for a new domain, the need that is another drawback in ALPNET.

Appendix

For the entry words in Examples [8] and [9], as sub-entries the second level of the xml tree for each entry includes the parts of speech in the entry polysynthetic word. To each sub-entry word two tags are assigned: an affixation tag, namely prefix, base, or suffix, and a POS tag, namely verb, noun, or particle. The sub-sub-entries at the next level of the xml tree are the attributes assigned to each word. For the base word there is another level of the tree where the word has been further tokenized into its affixes and stem tagged with its root, template, and structure.

Table 5 Tagset for Mobin

Constraint	Attribute (abbreviation) (عربی)	Values (abbreviation) (عربی)
Entry word	Affixation (Afix)	Prefix (Pref) Suffix (Suf) Circumfix (Circum) Base
	Part of speech (POS)	Verb (فعل) Noun (اسم) Particle (حرف)
	Declination (Decl)	Declined (Dc) (معرب) Non-declined (NDc) (مبنی)
Common for verb and noun	Alternation (Alt)	Salim (Slim) (سالم) Mahmuz (Mah) (مهموز) Muza"af (Muz) (مضاعف) Ajwaf (Ajv) (اجوف) Naghis (Nag) (ناقص) Lafif (Laf) (لفيف) ...
	Generation (Genr)	Generative (Genrt) (متصرف) Non-generative (NGenrt) (غير متصرف)
	Person (Prsn)	Speaker (Spkr) (متكلم) Addressee (Adrs) (مخاطب) Non-addressee (NAdrs) (غايب)
	Number (Num)	Singular (Sing) (مفرد) Dual (مثنی) Plural (Plr) (جمع)
	Gender (Gend)	Masculine (Masc) (مذكر) Feminine (Fem) (مؤنث) Neutral (Nut) (خنثی)
	Case	Nominative (Nomn) (مرفوع) Accusative (Acus) (منصوب) Genitive (Gent) (مجرور)
	Structure (Struct)	Simple trilateral (Simp3) (ثلاثی مجرد) Simple quadrilateral (Simp4) (رباعی مجرد) Simple pentalateral (Simp5) (خماسی مجرد) Compound trilateral (Comp3) (ثلاثی مزيد) Compound quadrilateral (Comp4) (رباعی مزيد) Compound pentalateral (Comp5) (خماسی مزيد)
	Template (Temp)	Hundreds of templatic patterns
	Root	Thousands of consonantal strands
	Stem extra particle (SEP)	

(continued)

Table 5 (continued)

Constraint	Attribute (abbreviation) (عربی)	Values (abbreviation) (عربی)
Verb	Mood (Mod)	Imperative (Imper) (امر) Indicative (Indv) Subjunctive (Subjn) Jussive (Juss)
	Voice (Voic)	Active (Act) (معلوم) Passive (Psi) (مجهول)
	Aspect	Perfective (Prf) Imperfective (ImPrf)
	Time	Past (ماضی) Present (Prst) (مضارع) Future (Fut)
Noun	Derivation (Derv)	Derived (Dervd) (مشتق) Non-derived (NDervd) (جامد) Null (Nul)
	Derivation type (DervT)	Agent noun (AgntN) (اسم فاعل) Patient noun (PatN) (اسم مفعول) Instrument noun (InstN) (اسم آلت) Time noun (TimeN) (اسم زمان) Place noun (اسم مکان) Similar quality (صفت مشبه) Exaggerative (اسم مبالغه) Comparative (اسم تفضیل) ...
Particle	Specificity (Spc)	Noun-specific (NSp) Verb-specific (VSp) Common
	Operation (Opr)	Accusativity (Acusy) (نصب) Genitivity (Genty) (جر) Jussitivity (Jusy) Neutral (Nut)
	Semantic (Sem)	Genitive particle (GentP) (اسم جر) Semi-verb (SemiV) (مشبه بالفعل) Exclusion (Exls) Future (Fut) Negative (Neg) ...

Example [8]

```
<Phr Ent="فَسَيَكْفِيكَهُمُ">
   <Word Ent="فَ" Afix="Pref" POS="Particle">
      <PAtrib Decl="NDc" Spc="NVCom" Opr="Nut" Sem="Adt">
      </PAtrib>
   </Word>
   <Word Ent="سَ" Afix="Pref" POS="Particle">
      <PAtrib Decl="NDc" Spc="VSp" Opr="Nut" Sem="Fut">
      </PAtrib>
   </Word>
   <Word Ent="يَكْفِي" Afix="Base" POS="Verb">
      <VAtrib Decl="Dc" Alt="Nag" Aspect="ImPrf" Mod="Moz" Time="Futr" Genr="Genrt"
        Prsn="NAdrse" Num="Sing" Gend="Masc" Case="Indv" Voic="Act">
      </VAtrib>
      <Morph>
         <Pref Ent="يَ" ISem="InflP">
         </Pref>
         <Stem Ent="كْفِي" Struct="Simp3" Temp="yaF~AaL" Root="ك+ف+و|ي" SEP="">
         </Stem>
      </Morph>
   </Word>
   <Word Ent="كَ" Afix="Suf" POS="Noun">
      <NAtrib Decl="NDc" DeclT="Pron" Alt="Nul" Prsn="Adrse" Num="Sing" Gend="Masc"
        Genr="NGenrt" Varied="Nul" Def="Deft" Case="Acus" Derv="Nul" DervT="Nul" Func="Pat">
      </NAtrib>
   </Word>
   <Word Ent="هُمُ" Afix="Suf" POS="Noun">
      <NAtrib Decl="NDc" DeclT="Pron" Alt="Nul" Prsn="NAdrse" Num="Plr" Gend="Masc"
        Genr="NGenrt" Varied="Nul" Def="Deft" Case="Acus" Derv="Nul" DervT="Nul" Func="Pat">
      </NAtrib>
   </Word>
</Phr>
```

Example [9]

```
   <Phr Ent="فَبِالْبَاطِلِ">
      <Word Ent="فَ" Afix="Pref" POS="Particle">
         <PAtrib Decl="NDc" Spc="NVCom" Opr="Nut" Sem="Adt">
         </PAtrib>
      </Word>
      <Word Ent="بِ" Afix="Pref" POS="Particle">
         <PAtrib Decl="NDc" Spc="NSp" Opr="Genty" Sem="GentP">
         </PAtrib>
      </Word>
      <Word Ent="الْ" Afix="Pref" POS="Noun">
         <NAtrib Decl="NDc" DeclT="Nut" Alt="Nul" Prsn="Nul" Num="Sing" Gend="Nut"
           Genr="NGenrt" Varied="Nul" Def="Deft" Case="Nul" Derv="Nul" DervT="Nul" Func="Nul">
         </NAtrib>
      </Word>
      <Word Ent="بَاطِلِ" Afix="Base" POS="Noun">
         <NAtrib Decl="Dc" DeclT="Nut" Alt="Slim" Prsn="Nul" Num="Sing" Gend="Masc"
           Genr="Genrt" Varied="Vri" Def="Deft" Case="Gent" Derv="Dervd" DervT="AgntN"
           Func="GentToG">
         </NAtrib>
         <Morph>
            <Stem Ent="بَاطِل" Struct="Comp3" Temp="Faa:AiL" Root="ب+ط+ل" SEP="ا">
            </Stem>
            <Suf Ent="ِ" ISem="Vowel">
            </Suf>
         </Morph>
      </Word>
   </Phr>
```

References

Al-Othman A, Al-Salman AM (2020) Arabic morphological analysis techniques: a survey and classification. Int J Adv Comp Sci Appl 11(2)

Al-Shartuti R (1992) Mabadi al-Arabiyah. Asatir Press, Tehran

Al-Sughaiyer IA, Al-Kharashi IA (2004) Arabic morphological analysis techniques: a comprehensive survey. J Am Soc Inform Sci Technol 55(3):189–213

Beesley KR (1996a) Arabic finite-state morphological analysis and generation. In: Proceedings of the 16th international conference on computational linguistics (COLING 1996). Copenhagen, Denmark

Benedetto D, Caglioti E, Loreto V (2002) Language trees and zipping. Phys Rev Lett 88(4)

Beesley KR (1996b) Arabic finite-state morphological analysis and generation. In: Proceedings of the 16th international conference on computational linguistics (COLING 1996). Copenhagen, Denmark

Bohas G, Guillaume J-P, Kouloughli DE (1990) The Arabic linguistic tradition. Routledge, London and New York

Buckwalter T (2004) Buckwalter Arabic morphological analyser Version 2. Liguistic Data Consortium

el-Dahdah A (1999) A dictionary of arabic grammar in charts and tables. Librairie du Liban Publishers, Beirut

Dichy J, Farghaly A (2007) Grammar-Lexis relations in the computational morphology of Arabic. In: Soudi et al (eds) Arabic computational morphology: knowledge-based and empirical methods. Springer, Dordrecht, the Netherlands

Dror J, Shaharabani D, Talmon R, Wintner S (2004) Morphological analysis of the Quran. Literary Linguistic Comput 19(4):431–452

Guillaume J-P (1988) Le Discours tout entier est nom, verbe et particle. Langages 92

Habash N, Rambow O, Roth R (2009) MADA+TOKAN: a toolkit for arabic tokenization, morphological disambiguation, POS tagging, stemming and lemmatization. In: Proceedings of the second international conference on arabic language resources and tools. The MEDAR Consortium

Karttunen L Beesley KR (2001) A short history of two-level morphology. Workshop at ESSLLI. University of Helsinki, Finland

Katamba F (1993) Morphology. MacMillan Press LTD.

Khorramshahi B (1994) Quranology: a collection of essays on Quranic topics (in Farsi). Mashreq Press, Tehran, Iran

Khorramshahi B (1999) The Encyclopedia of the holy Quran (in Farsi). Doostan Press, Tehran, Iran

McCarthy JA (1981) Prosodic morphology of non-concatenative morphology. In: Linguistic inquiry, vol 12, pp 373–418

Muhammadi H (2002) Zabane Quran. Dar al-Dhekr Press, Qom

Roth R, Rambow O, Habash N, Diab M, Rudin C (2008) Arabic morphological tagging, diacritization, and lemmatization using lexeme models and feature ranking. In: Proceedings of ACL, Columbus, OH

Shannon CE (1948) Bell Systems. Tech J 27:379

Shokrollahi-Far M, Hoseini S, Isazadeh A (2007a) Nobi: tagged corpus of Quran. In: The meeting of computational linguistics in the Netherlands, Leuven, Belgium

Shokrollahi-Far M, Isazadeh A, Barzegar I, Soltani R (2007b) Knowledge-base on Quran for tagging Arabic verbs. In: Proceedings of 1st international conference on digital communications and computer applications (DCCA), Jordan

Shokrollahi-Far M, Saraei J (2007) An enhanced knowledge-based tagger for Arabic verbs. In: Computational linguistics in the Netherlands (CLIN 18), Nijmegen, the Netherlands

Shokrollahi-Far M, Minaie B, Barzegar I, Hossein-Zadeh H, Ghasdi M, Hoseini S (2009) Bootstrapping tagged Islamic Corpora. In: Proceedings of international conference Arabic language resources and tools. Cairo, Egypt

Shokrollahi-Far M (2012) Mobin: a knowledge-based morpho-syntactic parser for Arabic. In: Proceedings of 16th CSI international symposium on artificial intelligence and signal processing. Shiraz, Iran.

Soudi A, van den Bosch A, Neuman G (2007) Arabic computational morphology: knowledge-based and empirical methods. Springer, Dordrecht, the Netherlands

Sproat R (1992) Morphology and computation. ACL-MIT Press

Versteegh K (1997) Landmarks in linguistic thought III: the arabic linguistic tradition. Routledge, London and New York

Wintner S (2007) Strengths and weaknesses of finite-state technology: a case study in morphological grammar development. Nat Lang Eng 14(4):457–469

Index

G. Y. Georgiev, M. Shokrollahi-Far (eds.), *Efficiency in Complex Systems*,
Springer Proceedings in Complexity, https://doi.org/10.1007/978-3-030-69288-9